Soils and Quaternary Geology
of the
Southwestern United States

Edited by

David L. Weide
Department of Geoscience
University of Nevada, Las Vegas
4505 Maryland Parkway
Las Vegas, Nevada 89154

Associate Technical Editor

Marianne L. Faber
Environmental Programs
Lockheed Engineering and
Management Services Company, Inc.
Las Vegas, Nevada 89114

SPECIAL PAPER
203

Published by The Geological Society of America, Inc.
3300 Penrose Place, P.O. Box 9140, Boulder, Colorado 80301

Printed in U.S.A.

Library of Congress Cataloging in Publication Data
Main entry under title:

Soils and Quaternary geology of the southwestern
    United States.

    (Special paper / Geological Society of America ; 203)
    Papers developed out of a symposium organized for the
1982 meeting of the Cordilleran Section of the Geological
Society of America, at Anaheim, Calif.
    Includes bibliographies and index.
    1. Geology, Stratigraphic—Quaternary—Congresses.
2. Soils—Southwestern States—Congresses. 3. Geology—)
Southwestern States—Congresses. I. Weide, David L.
II. Geological Society of America. Cordilleran Section.
III. Series: Special paper (Geological Society of America)
; 203.

QE696.S578   1985   552'.5   85-10001
ISBN 0-8137-2203-9

Cover photograph by Roger B. Morrison

Contents

Denis E. Marchand and son André, July 1974.

The late Denis Marchand was a resource of new ideas, a leader in Quaternary geology, and a dependable colleague to those who sought knowledge, advice, and interchange. This volume of papers is dedicated to him for his contributions and inspirations to the scientific community. Denny began his interdisciplinary approach to geologic research with his dissertation on the interactions of rock weathering, soil formation, and botanical associations in the White Mountains of California. His analysis of chemical transformations in geologic and biologic systems remains a landmark study using an unprecedented approach. Denny continued his innovative research by using soils and soil mapping to resolve geologic problems relating to late Cenozoic history, stratigraphy, and geomorphic processes in central Pennsylvania and central California. His most recent research on using soils and soil development to correlate and date geologic deposits is unequaled in breadth, scope, and detail. Perhaps most impressive was Denny's willingness to communicate, and his many conversations, lectures, and publications are treasured as archives by his friends and colleagues. Today, we can only aspire to advance our science in Denny's fashion of enthusiasm and curiosity for knowledge.

Jennifer Harden
Menlo Park, California
April 1984

Preface

During the past 15 years, geologists have become increasingly involved with the problem of how to date recent sedimentary deposits and tectonic activity. This exploration into the Holocene and late Pleistocene has been fuled by an increasing awareness of the essential role played by neotectonics and paleocolimates in guiding urban expansion and in planning hazardous-waste facilities and nuclear generating stations. One of the most useful tools for investigating the recent history of climate and faulting is pedology—the study of soil.

The papers in this volume present laboratory and field studies aimed at dating soils and determining the rate of soil-forming processes. The selection presented here grew out of a symposium organized for the 1982 meeting of the Cordilleran Section of the Geological Society of America, at Anaheim, California. Eight of the 15 papers presented at the meeting were expanded, integrated, and extensively rewritten to systematically explore soil chronology and pedogenic processes.

The volume is divided into two parts: the first group of five papers explores the pedogenic process of carbonate and clay accumulation and their rates and depths of formation. In addition, field studies described in the first group of papers show how soils may be used to date surfical deposits and recent movement on major faults. The second group of papers is a series of regional studies focusing on Quaternary alluvial stratigraphy, soils, landscape ages, and recent tectonic history in those areas of southwestern United States where an expanding population will eventually result in urbanization within zones of potential seismic activity.

David L. Weide
Las Vegas, Nevada
April 1984

Geological Society of America
Special Paper 203
1985

Calcic soils of the southwestern United States

Michael N. Machette
U.S. Geological Survey
Denver Federal Center
Denver, Colorado 80225

ABSTRACT

Calcic soils are commonly developed in Quaternary sediments throughout the arid and semiarid parts of the southwestern United States. In alluvial chronosequences, these soils have regional variations in their content of secondary calcium carbonate ($CaCO_3$) because of (1) the combined effects of the age of the soil, (2) the amount, seasonal distribution, and concentration of Ca^{++} in rainfall, and (3) the $CaCO_3$ content and net influx of airborne dust, silt, and sand. This study shows that the morphology and amount of secondary $CaCO_3$ (cS) are valuable correlation tools that can also be used to date calcic soils.

The structures in calcic soils are clues to their age and dissolution-precipitation history. Two additional stages of carbonate morphology, which are more advanced than the four stages previously described, are commonly formed in middle Pleistocene and older soils. Stage V morphology includes thick laminae and incipient pisolites, whereas Stage VI morphology includes the products of multiple cycles of brecciation, pisolite formation, and wholesale relamination of breccia fragments. Calcic soils that have Stage VI morphology are associated with the late(?) Miocene constructional surface of the Ogallala Formation of eastern New Mexico and western Texas and the early(?) Pliocene Mormon Mesa surface of the Muddy Creek Formation east of Las Vegas, Nevada. Thus, calcic soils can represent millions of years of formation and, in many cases, provide evidence of climatic, sedimentologic, and geologic events not otherwise recorded.

The whole-profile secondary $CaCO_3$ content (cS) is a powerful developmental index for calcic soils: cS is defined as the weight of $CaCO_3$ in a 1-cm^2 vertical column through the soil (g/cm^2). This value is calculated from the thickness, $CaCO_3$ concentration, and bulk density of calcic horizons in the soil. (See Soil Survey Staff, 1975, p. 45–46, for a complete definition of calcic horizon.) $CaCO_3$ precipitates in the soil through leaching of external Ca^{++} that is deposited on the surface and in the upper part of the soil, generally in the A and B horizons. The cS content, maximum stage of $CaCO_3$ morphology, and accumulation rate of $CaCO_3$ in calcic soils of equivalent age can vary over large regions of the southwestern United States in response to regional climatic patterns and the influx of Ca^{++} dissolved in rainwater and solid $CaCO_3$.

Preliminary uranium-trend ages and cS contents for relict soils of the Las Cruces, New Mexico, chronosequence show that 100,000- to 500,000-year-old soils have similar average rates of $CaCO_3$ accumulation. Conversely, soils formed during the past 50,000 years have accumulated $CaCO_3$ about twice as fast, probably because the amount of vegetative cover decreased in the Holocene and, hence, the potential supply of airborne Ca^{++} and $CaCO_3$ to the soil surface increased.

The quantitative soil-development index cS can be used to estimate the age of calcic soils. This index can also be used to correlate soils formed in unconsolidated Quaternary sediments both locally and regionally, to compare rates of secondary $CaCO_3$ accumulation, and to study landscape evolution as it applies to problems such as earthquake hazards and siting of critical facilities.

M. N. Machette

INTRODUCTION

Quantitative indices of soil development are powerful and fundamental tools for estimating ages of Quaternary soils, but such indices have not been widely applied in geology as yet. However, many soils and their parent materials can scarcely be dated otherwise and thus, inadequate age control limits the value of many geologic investigations. This paper describes a quantitative index that assesses the development of calcic soils: that is, those arid and semiarid soils that have significant accumulations of secondary calcium carbonate. This index can be used to (1) estimate the duration of soil formation, (2) correlate Quaternary sediments in which they are formed, and (3) analyze spatial and temporal variations in rates of secondary $CaCO_3$ accumulation.

Calcic soils are herein defined as having significant accumulations of secondary calcium carbonate. [Note: The term carbonate is used hereafter for both calcium carbonate and magnesium carbonate, which may occur as a minor constituent in calcic soils.] According to pedologic criteria they must have a calcic horizon. In the field, such soils are characterized by a layer (at least 15 cm thick) of secondary carbonate that is enriched in comparison to the soil's parent material.

If Quaternary sediments or soils contain appropriate material, they can be dated by one of several analytical methods. Organic or inorganic carbon can be dated by C^{14} or uranium-disequilibrium methods. Ages of volcanic rocks can be dated by the K-Ar method, and their associated tephra (ash and pumice) can also be dated by the fission-track technique or correlated using chemical and mineralogical characteristics (tephrochronology). These techniques, especially when used with stratigraphic and geomorphic evidence, help solve many problems in Quaternary geology. Because of advances in dating techniques, age control in many Quaternary geologic studies has been much improved. Unfortunately, most Quaternary sediments lack abundant datable material and thus their stratigraphies are commonly based on less exacting, yet widely applicable criteria.

Soils, however, mantle most land surfaces. These soils range from thin, incipient to weak profiles on young (Holocene) alluvium, to thick, strong profiles on old (Pleistocene and Pliocene) alluvium. The vast majority of soils, however, are between these extremes. Calcic soils are at the arid end-member of a broad spectrum of climates under which soils form. This study deals only with relict soils (as redefined by Ruhe, 1965, and as used by Birkeland, 1974). Relict soils are those that have formed continuously since their parent material was deposited. The age of a relict soil should closely approximate that of its parent material because the soil is the cumulative product of soil formation since the deposit became geomorphically stable, that is isolated from deposition or erosion.

Calcic soils are particularly well suited to quantitative assessments of soil development. In the United States, these soils are widespread in the Southwest, but they also extend into the northern Basin and Range province and into the Midwest (Fig. 1).

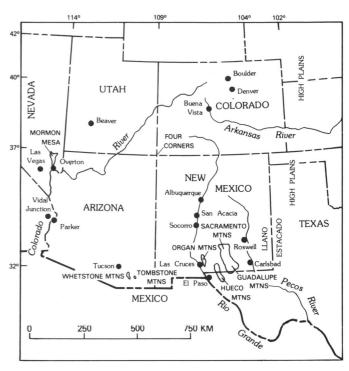

Figure 1. Geographic and cultural features of the southwestern United States mentioned in this report.

They form widely in unconsolidated sediments during periods ranging from as little as a few thousand years to as much as several million years. Most calcic soils in the Southwest have formed since middle and late Pleistocene time, but some highly advanced ones have been forming since Pliocene or late Miocene time on stable surfaces such as the Mormon Mesa of southern Nevada and the High Plains of Colorado, New Mexico, and Texas.

This report is based on reconnaissance and detailed soil studies in the southwestern United States during 1974 through 1977 under contract to the U.S. Nuclear Regulatory Commission. Preliminary results were published by Bachman and Machette in 1977. Since then, under numerous U.S. Geological Survey programs, further information has been collected in New Mexico and in the Basin and Range province near Beaver, Utah (Fig. 1; Machette, 1982).

NOMENCLATURE AND DISTRIBUTION OF CALCIC SOIL AND PEDOGENIC CALCRETE

Nomenclature

The term "caliche" (Blake, 1902) has been broadly applied to deposits of secondary calcium carbonate of various origins, often irrespective of their geomorphic and stratigraphic relations, physical and chemical characteristics, and genesis. Although geologists in the United States generally associate caliche with soil carbonate, the indiscriminate use of the term has created semantic problems. For example, the term has been applied loosely to the soluble salts of sodium in nitrate deposits of northern Chile and Peru and to gypsum deposits in Death Valley, California (cited in Goudie, 1973, p. 8–9). Such usage can lead to misconceptions, especially if comparisons and interpretations are made in either a stratigraphic or a genetic context. In general, the term caliche should be abandoned for more specific nomenclature.

Calcic soils (an informal term) are herein defined as those soils that contain a significant amount of secondary carbonate, generally in the form of calcic horizon(s). The term "horizon" is used to describe a three-dimensional body (a layer) of soil material. For the index developed here, such horizons should represent new additions of secondary carbonate to a soil profile, rather than redistributions of primary (allogenic) carbonate. Calcic horizons may range from layers of carbonate-coated pebbles or carbonate filaments and nodules to thick, massive, and indurated carbonate-enriched layers. With increasing age, calcic horizons become so well developed that their structural morphology and general appearance are dominated by the presence of carbonate.

The nongenetic term "calcrete," introduced by Lamplugh (1902) and popularized by Goudie (1973), refers to indurated masses of calcium carbonate. Calcrete is herein restricted to near-surface or shallow, terrestrial deposits of calcium carbonate that have accumulated in or replaced a preexisting soil, unconsolidated deposit, or weathered rock material, to produce an indurated mass (definition slightly modified from that of Goudie, 1973, p. 5). Following this usage, calcretes include indurated calcic soils (pedogenic calcrete) as well as lacustrine marls (lacustrine calcretes), ground-water deposits (ground-water calcretes), and spring or travertine deposits. Some calcic soils may result from a combination of pedogenic and nonpedogenic processes and their respective components may be virtually indistinguishable. For example, a ground-water calcrete that is exhumed and then modified by pedogenesis may form many of the same morphologic features commonly seen in pedogenic calcretes.

In the field, I use the terms "calcic soil" and "pedogenic calcrete" instead of "caliche" to denote those calcium-carbonate-enriched deposits that are formed exclusively by soil processes (pedogenesis). "Calcic soils" is a useful collective term that is roughly synonymous with the Pedocals of Marbut's (1935) soil classification; although many geologists favor the term pedocal (Birkeland, 1984), it has fallen from current pedologic usage.

Pedologists commonly recognize three master horizons in soils of arid and semiarid climates: the A, B, and C horizons, with increasing depth. Horizons with secondary carbonate are denoted by a "k" suffix, as in the Bk horizon (Guthrie and Witty 1982; Birkeland, 1984). Although this usage is consistent with modifiers used to describe other horizon types, it gives no indication of the amount of carbonate accumulation. Because calcic soils are characterized by carbonate, nomenclature that indicates the amount of carbonate accumulation would be useful for geologic purposes.

Gile and others (1965) recognized the need for descriptive nomenclature based on field criteria that would better describe calcic soils. They introduced the term "K horizon," a master soil horizon, to describe prominent layers of carbonate accumulation. The K-horizon nomenclature is particularly useful for geologic applications because it allows field differentiation of weak calcic horizons, such as Bk horizons, from moderate to strong K horizons. According to their definition, the K horizon must have more than 90 percent K-fabric, a diagnostic soil fabric in which ". . . fine-grained authigenic carbonate occurs as an essentially continuous medium. It coats or engulfs, and commonly separates and cements skeletal pebbles, sand, and silt grains . . ." (Gile and others, 1965, p. 74).

Distribution

Calcic soils are found in many areas of the western United States where warm and periodically to predominately dry climates result in torric, ustic, or xeric soil-moisture regimes (see Soil Survey Staff, 1975, for descriptions of soil-moisture regimes). Most calcic soils form under grassland or desert-type vegetation, but some may exist under pinon-pine and juniper forest vegetation where the present soil-moisture conditions differ from those that existed when carbonate was accumulating in the soil. Figure 2 shows major areas of calcic soils in the western United States as interpreted from the United States Soils Map (U.S. Soil Conservation Service, 1970). The main group of calcic soils includes only those soils that (1) are in the Entisol, Mollisol, Alfisol, or Aridisol orders (Soil Survey Staff, 1975); (2) have calcic or petrocalcic horizons; and (3) have ustic or torric soil-moisture regimes. Soils having xeric regimes receive mainly winter rainfall and are associated with Mediterranean climates such as that in California's San Joaquin Valley (Fig. 2). Most areas having these climates are too moist for soil $CaCO_3$ to accumulate; thus they are shown as marginal areas of accumulation (Fig. 2). However, soils formed in calcareous bedrock (such as limestone or dolomite) or alluvium are not included in Figure 2 because nonpedogenic accumulations of carbonate can form in these areas under almost any climate or can result strictly from high levels of ground water.

Within the United States, soils that continually accumulate carbonate generally are restricted to areas of arid and semiarid climate within low-altitude basins of the Southwest, although some soils accumulate carbonate periodically at higher altitudes in the northern Basin and Range province and in the lower altitudes of mountainous regions such as the Colorado Plateaus province. Gile (1975, 1977) has demonstrated that the increased rainfall associated with climatic gradients near mountain ranges has a profound effect on the distribution and concentration of carbonate in Holocene soils of the Las Cruces, New Mexico, area (Fig. 1). In the Southwest, Bk or K horizons formed in a single-age deposit generally become less calcareous or less continuous along soil transects from basins to mountains.

The boundary between soil that has accumulated carbonate and soil that is leached of carbonate is called the pedocal-pedalfer boundary. Jenny (1941) demonstrated that the pedocal-pedalfer boundary in the Midwest probably is controlled by the eastward

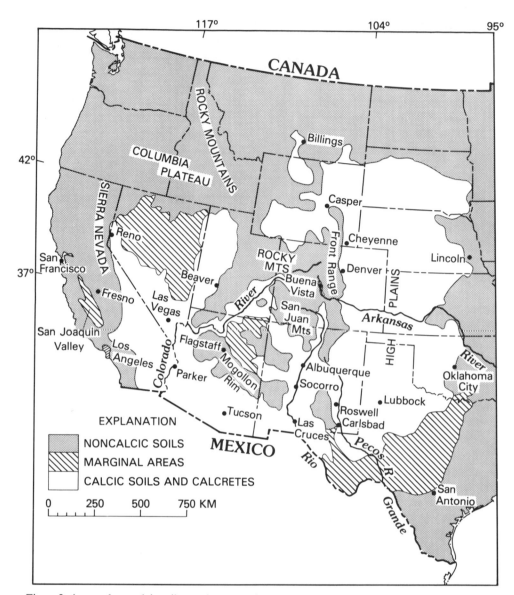

Figure 2. Areas where calcic soils may be present in the southwestern United States, as interpreted from the soils map of the United States (U.S. Soil Conservation Service, 1970). Areas of discontinuous or poorly preserved calcic soils are designated marginal.

increase in soil moisture, vegetative cover, soil organic-matter content, and soil acidity (decrease in soil pH). These factors, in combination, enable more carbonate to be leached from a soil than is supplied to its surface; the result is a pedalfer, a soil leached of carbonate. The eastern boundary of calcic soils shown in Figure 2, which I interpreted from the United States Soils Map, is similar to that of Jenny's (1941, Fig. 99).

MORPHOLOGY OF CALCIC SOILS

Gile and others (1966) described a morphological sequence for calcic soils in arid and semiarid regions of New Mexico based on the physical characteristics of pedogenic calcium carbonate. This sequence includes four stages of morphology (soil structure)

that depend partly on the texture of a soil's parent material. Bachman and Machette's (1977) soil studies in the Southwest led to the recognition of two additional, more advanced stages of carbonate morphology (Stages V and VI, Table 1; modified from Bachman and Machette, 1977, Table 2). These two stages are characterized by thick laminae, pisolites, and multiple episodes of brecciation (physical disaggregation) and recementation. The diagnostic characteristics used to distinguish the six stages of carbonate morphology are shown in Table 1.

Gile and others (1965) proposed Stage IV as the maximum degree of carbonate morphology and included all thicknesses of laminae within this stage. However, during Bachman and Machette's reconnaissance, they recognized more advanced morphology in old calcic soils, particularly in pedogenic calcretes.

TABLE 1. STAGES OF CALCIUM CARBONATE MORPHOLOGY OBSERVED IN CALCIC SOILS AND PEDOGENIC CALCRETES
DEVELOPED IN NONCALCAREOUS PARENT MATERIALS UNDER ARID AND SEMIARID CLIMATES OF THE AMERICAN SOUTHWEST
[Modified from Gile and others (1966, p. 348, Table 1) and Bachman and Machette (1977, p. 40, Tables 2 and 3)]

Stage	Gravel content[1]	Diagnostic morphologic characteristics	CaCO₃ distribution	Maximum CaCO₃ content[2]
		CALCIC SOILS		
I	High-----	Thin, discontinuous coatings on pebbles, usually on undersides.	Coatings sparse to common	Tr-2
	Low------	A few filaments in soil or faint coatings on ped faces.	Filaments sparse to common	Tr-4
II	High-----	Continuous, thin to thick coatings on tops and undersides of pebbles.	Coatings common, some carbonate in matrix, but matrix still loose.	2-10
	Low------	Nodules, soft, 0.5 cm to 4 cm in diameter	Nodules common, matrix generally noncalcareous to slightly calcareous.	4-20
III	High-----	Massive accumulations between clasts, becomes cemented in advanced form.	Essentially continuous dispersion in matrix (K fabric).	10-25
	Low------	Many coalesced nodules, matrix is firmly to moderately cemented.	------------------do---------------	20-60
		PEDOGENIC CALCRETES (INDURATED CALCIC SOILS)		
IV	Any------	Thin (<0.2 cm) to moderately thick (1 cm) laminae in upper part of Km horizon. Thin laminae may drape over fractured surfaces	Cemented platy to weak tabular structure and indurated laminae. Km horizon is 0.5-1 m thick.	>25 in high gravel content >60 in low gravel content
V	Any------	Thick laminae (>1 cm) and thin to thick pisolites. Vertical faces and fractures are coated with laminated carbonate (case-hardened surface)	Indurated dense, strong platy to tabular structure. Km horizon is 1-2 m thick.	>50 in high gravel content >75 in low gravel content
VI	Any------	Multiple generations of laminae, breccia, and pisolites; recemented. Many case-hardened surfaces.	Indurated and dense, thick strong tabular structure. Km horizon is commonly >2 m thick.	>75 in all gravel contents

[1]High is more than 50 percent gravel; low is less than 20 percent gravel.
[2]Percent CaCO₃ in the <2-mm-fraction of the soil. Tr, trace of carbonate.

Therefore, I propose that Stage IV morphology be limited to K horizons that have laminae or laminar layers less than 1 cm thick. The next stage of morphology, Stage V (Fig. 3A), is distinguished by laminae or laminar layers thicker than 1 cm and, in some cases, by incipient to thick, concentrically banded pisolitic structures. Pisoliths commonly have cores of fragments of Stage III soil matrix or Stage IV laminae, and these fragments suggest that brecciation is an integral part of pisolith formation in soils.

Calcic soils of Stage VI morphology are characterized by multiple episodes of brecciation and pisolith formation through relamination and recementation of breccia fragments (Figure 3B); the resulting products are thick, indurated masses of carbonate (pedogenic calcretes). These soils are the climax products of relatively continuous carbonate accumulation over perhaps millions of years. In eastern New Mexico, Stage V calcic soils have formed in middle to early Pleistocene alluvium that contains reworked clasts of older calcic soils (Stage VI) of the Miocene Ogallala Formation, indicating that the processes of soil brecciation and pisolization were active throughout the Quaternary, and possibly during the Pliocene.

Calcic soils that have Stage V and VI morphology result from varied conditions of carbonate accumulation during their prolonged formation. Bretz and Horberg (1949) considered calcic soils with Stage V and VI morphology to be self-brecciation features and named them collectively "Rock House structures" for outcrops near Rock House, New Mexico. Bryan and Albritton (1943) indicated that these structures are related to fluctuations in soil moisture caused by alternating climatic conditions, probably fluctuations between cool, moist conditions of the pluvials and warm, dry conditions of the interpluvials.

Stage V and VI calcic soils are particularly well preserved at numerous locations in the Southwest. The most spectacular of these soils are associated with the constructional geomorphic surfaces of the following geologic units:

CaCO₃ stage	Geomorphic surface and probable age	Geologic unit and probable age	Location
V	Upper La Mesa, early Pleistocene	Camp Rice Formation of Strain (1966), middle Pleistocene to Pliocene	Las Cruces, New Mexico
V	Mescalero middle Pleistocene	Gatuna Formation, middle Pleistocene	Southeastern New Mexico
VI	Mormon Mesa, early(?) Pliocene	Muddy Creek Formation, Miocene	Overton, Nevada
VI	Ogallala, late(?) Miocene	Ogallala Formation, Miocene	Eastern New Mexico and western Texas

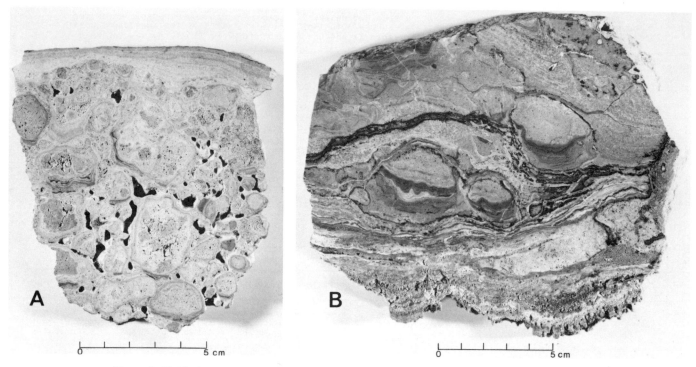

Figure 3. Slabbed samples of calcic soils showing advanced stages of carbonate morphology. A, Incipient Stage V morphology found in the upper part (K21m subhorizon) of the pedogenic calcrete of the upper La Mesa surface (early Pleistocene) near Las Cruces, New Mexico (U.S. Soil Conservation Service pedon 60-NMex7-7). B, Stage VI morphology typically found in the upper part of the Ogallala caprock (late? Miocene) in eastern New Mexico and western Texas.

Gile and others (1966) have shown that the difference in texture between gravelly and nongravelly parent materials greatly influences the time required to form Stage I, II, III, and IV morphologies (Table 1). However, parent-material texture is less significant in the two ultimate stages (V and VI), because of the great expansion that accompanies their formation. Stage V and IV horizons have so much $CaCO_3$—commonly more than 50 percent in gravelly materials and more than 75 percent in fine-grained materials—that the texture of the parent material is completely obscured in these horizons.

Volumetric calculations of the carbonate and noncarbonate fractions of calcic soils by Gardner (1972) and Bachman and Machette (1977, Table 7) indicate that K horizons of strong Stage III or greater morphology undergo marked and progressive expansion with continuing accumulation of $CaCO_3$. If the volume of carbonate that accumulates in a soil exceeds the original pore space of the soil's parent material, there must be a physical expansion of the detrital grains. Viewed in thin section, these K horizons have scattered detrital grains that appear to float in a matrix of carbonate. For example, there must be 400 to 700 percent volumetric expansion in the original framework of detrital grains to accommodate the carbonate found in some K horizons of Stage V and VI morphology.

As calcic soils accumulate carbonate and exceed Stage III morphology, their bulk density systematically and progressively increases with a concomitant decrease in porosity and permeabil-ity. Figure 4 shows the relation between carbonate content and bulk density for pedogenic calcretes having Stage IV, V, or VI morphology. Over millions of years of carbonate accumulation, these soils obtain carbonate concentrations (in percent) and bulk densities that approach those of limestones (Fig. 4). These physi-cal characteristics and the gross similarity of laminar and pisolitic features to those produced by algal masses have led some investi-gators to consider the Ogallala caprock a lacustrine deposit rather than a calcic soil (compare Fig. 3B this report with Elias, 1931, pl. 21B; Price and others, 1946).

PROCESSES OF CALCIUM CARBONATE ACCUMULATION

Secondary carbonate may accumulate through several var-ied processes to form either calcic soils or other deposits that resemble soils. Although major processes leading to such concen-trations have been discussed by Goudie (1973), four of these processes are briefly reviewed here, because some calcic soils can easily be confused with strictly nonpedogenic accumulations of carbonate.

Many calcic soils were once thought to form by "upward capillary flow of calcareous water, induced by constant and rapid evaporation at the surface in a comparatively rainless region" (Blake, 1902, p. 225). In the Southwest, ascending $CaCO_3$-rich water can and should precipitate some $CaCO_3$. However, this

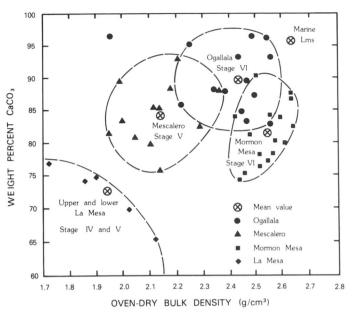

Figure 4. Percent calcium carbonate (whole-soil fraction) versus oven-dry bulk density (g/cm^3) of some pedogenic calcretes in the Southwest. Includes only data for K horizons of Stage IV, V, or VI morphology. The carbonate contents (Clarke, 1924) and bulk densities (Krynine and Judd, 1957) of marine limestones are also shown for comparison.

process is not likely to form calcic soils in areas of entrenched Pleistocene drainage, such as all along the Rio Grande in New Mexico. The ground water from which CaCO$_3$ might be derived has remained at levels well below the surface since the deposition of the soil parent material or shortly thereafter (Gile and others, 1981). Secondly, the concentration of Ca^{++} is usually low in ground water, thereby limiting the potential amount of carbonate that could be precipitated if ground water were to reach the surface and evaporate. Thirdly, in the Southwest many soils are formed in medium- to coarse-textured sediments that have little or no potential for capillary rise. Thus, the amount of rise postulated for this process requires an optimum combination of evaporative, ground-water level, and soil-texture conditions that does not commonly exist in the Southwest. The ascension process, as it is termed, does not appear to have contributed significant amounts of carbonate to most calcic soils of the Southwest.

A second process by which carbonate can accumulate involves the in situ weathering of Ca^{++} into soil water, and the subsequent precipitation of CaCO$_3$ in the soil. Ca^{++}-rich rocks such as basalt often are proposed as potential sources of Ca^{++} for this process. Such rocks, however, do not easily weather in semi-arid to arid environments, nor do they provide abundant Ca^{++} if they are weathered. Gardner (1972) evaluated the likelihood that this process could have formed the Mormon Mesa calcrete of southern Nevada and concluded that the concentration of carbonate in the calcrete would require complete removal of Ca^{++} from 100 m of basaltic gravel. Studies of the chemistry of similar calcic soils (Vanden Hueval, 1966; Aristarain, 1970; Bachman and Machette, 1977) found little evidence of depletion of rela-

tively mobile cations (such as Ca^{++}, Na^{++}, and K^{++}) in the detrital mineral fraction of the soil. These same soils lack residual accumulations of relatively immobile elements, such as Al, Si, or Ti, that should exist if the carbonate were concentrated by weathering. Thus, it seems that the process of in situ weathering is not involved in the formation of the majority of calcic soils in the Southwest. Obvious exceptions are soils formed on parent materials that contain limestone, calcareous sandstones, or dolomite. For example, the alluvium flanking the eastern side of the Sacramento and Guadalupe Mountains in southern New Mexico (Fig. 1) has soils formed partly by in situ weathering of limestones and partly by leaching of airborne calcareous dust (see following discussion of the per descensum model).

Laterally flowing, CaCO$_3$-rich ground water commonly forms deposits that are misidentified as calcic soils or pedogenic calcretes. This process calls for Ca^{++}-charged ground water to either discharge onto a stream bottom or reach a near-surface position where Ca^{++} is concentrated by evaporation. Supersaturation of Ca^{++} causes precipitation of CaCO$_3$ and subsequent cementation of relatively pervious sands and gravels. Such ground-water calcretes are typically well indurated to depths of 10 m or more, are characterized by gravel clasts that have grain-to-grain contact, and generally lack the horizonation and morphologic structures common in calcic soils. Ground-water calcretes form quickly but at differing times as the subsurface or surface flow shifts laterally into more permeable material. Surface runoff may add to or redistribute this same CaCO$_3$ and produce laminar zones that resemble pedogenic calcretes of Stage IV and V morphology. In southeastern New Mexico, Bachman and Machette (1977) found ground-water calcretes that had laminae as much as 1 cm thick along highway drainage culverts in limestone-rich alluvium. These laminae prove gulley-bed cementation can occur rapidly.

Lattman (1973) found that gully-bed cementation occurs during surface runoff on limestone-rich alluvial-fan sediments near Las Vegas, Nevada, and proposed that this process could form the laminar horizons (Stage IV and V) of many pedogenic calcretes. I disagree with his hypothesis, as it applies to most calcic soils, because laminar K horizons are commonly overlain by Bk horizons that are formed in material of the same age as the K horizon. However, my investigations of calcretes in the Las Vegas area suggest that many of them are pedogenically modified ground-water calcretes. Bachman and Machette (1977) found calcretes of similar origin west of Roswell and Carlsbad, New Mexico; near the Whetstone and Tombstone Mountains of southern Arizona; and south of the Hueco Mountains in west Texas (Fig. 1).

The fourth process, which involves airborne supply of calcareous material to the soil surface, has until recently not been widely accepted as an important formational process for calcic soils in semiarid and arid regions of the United States, partly because of the pervasive and subtle nature of airfall. Goudie (1973, p. 136) termed this the "per descensum model" because CaCO$_3$ is leached from the surface and upper horizons of the soil and subsequently precipitates and accumulates in lower soil horizons at a depth controlled by soil moisture and texture (see discussion by McFadden and Tinsley, this volume). The carbonate in the soil comes mainly from external sources such as minute amounts of solid carbonate in aerosolic dust, silt, and eolian sand, and Ca^{++} dissolved in rainwater (Gardner, 1972; Gile and others, 1979). Over thousands of years, the continual translocation of Ca^{++} and precipitation of $CaCO_3$ forms calcic horizons; over millions of years, the accumulation of carbonate forms thick, dense, indurated calcic soils (pedogenic calcretes). Almost a quarter of a century ago, Brown (1956, p. 14) recognized that the Ogallala caprock was pedogenic when he stated "the wind and rain bring in all the materials that form the soil and its associated caliche and deposit these materials on the soil surface."

I believe that airborne $CaCO_3$ and Ca^{++} dissolved in rainwater are the predominant sources of carbonate in calcic soils that have formed over thousands to millions of years in the Southwest. Four lines of evidence support this conclusion. (1) Many calcic soils are in noncalcareous sediments well above present and former levels of ground water. (2) A relict calcic soil's development is directly related to the age of the associated geomorphic surface (and parent material); this relation need not be true if the carbonate were derived from ground-water sources. (3) Local sources, such as calcareous alluvium or lacustrine sediment, and calcic soils themselves, provide abundant carbonate that is transported by the wind. Recent dust fall in the Las Cruces, New Mexico, area has averaged about 0.2 g of $CaCO_3$ per cm^2 per 1,000 years (Gile and others, 1979), which is only slightly less than the long-term average rate of carbonate accumulation in soils of this region (see section on "Rates of $CaCO_3$ Accumulation"). (4) Finally, the concentration of Ca^{++} in rainfall is high in the Southwest; values in the southern New Mexico area and the Four Corners region may exceed 5 mg of Ca^{++} per liter of water (Junge and Werby, 1958). The Ca^{++} in this rainwater alone would provide a major part of a soil's carbonate if all the rainwater entered the soil and the Ca^{++} was precipitated as $CaCO_3$. Gile and others (1979 and 1981) have been foremost in documenting these sources of soil carbonate and in popularizing the per descensum model for calcic soils of the Southwest.

ASSESSING THE DEVELOPMENT OF CALCIC SOILS AND THEIR CARBONATE CONTENT

Most assessments of soil development are based on systematic changes in soil properties in relation to those of the soil's parent material. Such assessments can be based on single properties such as soil structure, clay concentration, color, or horizon thickness, or on combinations of these soil properties. The soil-development index of Harden (1982) illustrates the importance of integrating multiple soil properties with soil thickness for quantitative assessments. Yet the results of even this powerful index are affected by variations in parent-material permeability, porosity, and texture, parameters that influence the distribution and concentration of many soil properties. To be widely applicable, an index must compensate for or equalize variations in parent-material characteristics.

Secondary Carbonate Content

This study and many past studies show that with age calcic horizons generally increase in degree of $CaCO_3$ morphology, in concentration of $CaCO_3$, and in thickness. But they also show that these horizons are strongly influenced by the character of the soil's parent material. For example, a calcic horizon in silt and another in gravel of the same age can have different stages of $CaCO_3$ morphology, different concentrations of $CaCO_3$, and different thicknesses, and therefore may appear quite dissimilar. For these reasons I believe that the whole-profile index of secondary carbonate is the best quantitative measure of calcic soil development.

Secondary carbonate (cS) is that component of the soil carbonate that has accumulated since deposition of the parent material. Any carbonate initially in the parent material is here considered to be a primary component (cP), even though some of this carbonate may be remobilized in the soil. The cS is a quantitative measure that integrates the concentration and distribution of carbonate throughout calcic horizons of the soil. For example, a soil in nongravelly sand could have 20 percent $CaCO_3$ (the secondary component) uniformly distributed over a ½-m thickness and still have the same cS content as a soil that has 50 percent gravel and 30 percent $CaCO_3$ (secondary component in the <2 mm fraction) uniformly distributed over a two-third-meter thickness. The soil in coarse-grained material appears stronger in outcrop, mainly because coarse sands and gravels have less surface area to coat with carbonate than do silts and clays. Therefore, by considering both the concentration and the distribution of secondary carbonate in a soil, the texture of a parent material can be controlled as a nonessential factor in Jenny's (1941) equation of soil formation.

Methodology

The whole-profile index of secondary carbonate (cS) is the difference between the soil's total carbonate content (cT) and the amount of primary carbonate (cP) initially present in the soil's parent material ($cS = cT - cP$). The total carbonate content of a single horizon or sample interval (ct) consists of two separate components: cs (secondary $CaCO_3$) and cp (primary $CaCO_3$), where $cs = ct - cp$. The sum of cs values is cS. Values of cs are computed for each sample interval (usually a soil subhorizon) from $CaCO_3$ content, thickness, and bulk density as follows: $cs = c_3 p_3 d_3 - c_1 p_1 d_1$ where,

c_3 is the present total $CaCO_3$ content (g $CaCO_3$/100 g oven-dry soil),

c_1 is the initial $CaCO_3$ content (g $CaCO_3$/100 g oven-dry soil),

p_3 is the present oven-dry bulk density (g/cm^3),

p_1 is the initial oven-dry bulk density (g/cm^3),

d_3 is the thickness (in cm) of the sampled interval, and

d_1 is the initial thickness (in cm) of the sampled interval.

Values for the soil's initial CaCO$_3$ content (c_1) and bulk density (p_1) have to be estimated. Although this requirement may seem to be a serious limitation to the technique, reasonable estimates of parent-material values can be made from unweathered and noncalcareous material (Cn horizons) exposed at the base of the soil. Also, because K horizons expand volumetrically, their initial thickness (d_1) should be less than their present thickness (d_3), even though one must use $d_3 = d_1$. This assumption will cause the cS values to be slightly overstated for some K horizons. A more complicated but realistic measure may be to use $d_1 = d_3 (p_1/p_3)$.

Values for cS, expressed in g/cm^2, are the weight of pure CaCO$_3$ in a 1-cm^2 column through all soil horizons that contain carbonate. Machette (1978b, Table 1 and Fig. 7) has an illustration of how cS values are calculated from laboratory and soil-description data for some surface and buried calcic soils. Note that the total CaCO$_3$ contents (cT) calculated by Gile, Hawley, and Grossman (1981, Table 25) for the Las Cruces soils are expressed in kg of CaCO$_3$/m^2, dimensional units that are 10 times larger than those discussed in this report.

Sampling Design

Jenny (1941) presented an equation for the formation of soil that was a function of five main factors: climate, landscape relief, biotic activity, parent material, and time. To relate soil development to soil age in his equation, one must control or equalize the four other factors of soil formation. Such control can be obtained by considering soils in a chronosequence, which Harden and Marchand (1977, p. 22) define as "a group of soils whose differing characteristics are primarily or entirely the result of differences in the age of the parent material from which they formed, the other soil-forming factors being held constant or nearly so." Calcic soils were sampled from chronosequences that extend over small geographic areas to control potential regional variations in the influx rates of airborne CaCO$_3$ and Ca^{++}. Most of the sampling sites have similar positions in the landscape, have similar amounts and types of vegetative cover, have low-carbonate parent materials, and are in areas of similar climate or have similar soil-moisture regimes. Through this sampling design, a calcic soil's age is directly related to its index of cS.

An assumption basic to this technique is that carbonate has continually accumulated in the soils of arid and semiarid environments. Some carbonate could have been periodically lost from the soil, because soil-moisture conditions probably varied during the Quaternary and late Pliocene. However, in the regions sampled, the climatic changes probably caused the zone of carbonate accumulation only to move deeper in the soil, not through the soil. If soils of these regions had lost carbonate through excessive leaching, one might expect to find the evidence as buried noncalcareous soils (pedalfers), but buried pedalfers are rare

in these regions. The progressive burial of middle through upper Pleistocene calcic soils along the Organ Mountains of southern New Mexico (Gile and Hawley, 1966) provides clear evidence of continuous carbonate accumulation in southern New Mexico during this time.

Sampling and Analytical Techniques

To calculate cS, one needs to describe the properties of the soil and to determine soil horizon and subhorizon boundaries from pedogenic and parent-material changes. I use the K-horizon nomenclature of Gile, Peterson, and Grossman (1965) and the six-stage morphologic sequence to describe soil properties and horizon structures. Horizons and subhorizons thicker than 20 to 25 cm are subdivided into thinner units for sampling and laboratory analysis. The maximum sampling depth is restricted only to the base of the lowest calcic horizon, because the depth of carbonate accumulation varies according to the texture of the parent material. Samples are collected from vertical channels, and care is taken to include representative amounts of soil matrix and gravel. Peds and fragments of indurated material are collected and separately packaged for laboratory determinations of bulk density. If the soil is friable or does not contain peds, a piston-type sampling tube is used to extract a prescribed volume of material; the bulk density is then determined from the sample's dry weight and volume. Sampling must extend completely through the soil and into the underlying noncalcareous parent material (Cn horizon), commonly to several meters depth, to ensure reasonable estimation of c_1 and p_1 values for the soil. Soils should be sampled more than once if they vary considerably over lateral exposures.

Samples are air dried in the laboratory, pulverized by hand using a ceramic mortar and rubber pestle, and sieved to determine the amount of <2-mm- and >2-mm-size material. Because they will not constitute part of the total CaCO$_3$ in the soil, gravel clasts that are not calcareous and that are not coated with carbonate are discarded after sieving and after their weight percentage of the whole soil has been determined. Conversely, if the clasts have calcareous coatings, these coatings must be removed with acid to determine their contribution to cT. Samples of indurated calcic horizons are pulverized in a rotating-plate crusher and their carbonate data are reported on a whole-sample basis.

The <2-mm fraction is oven-dried at 105°C and separated into several equal portions, and the CaCO$_3$ content of one or more of these portions is determined with the Chittick device using a gasometric CaCO$_3$-dissolution technique that requires 1 to 5 g of <2-mm material per analysis. The standard operating technique and analytical precision of the method are discussed by Dreimanis (1962). This method provides quick, efficient, and inexpensive determinations of carbonate content (Bachman and Machette, 1977, appendix).

Bulk density (in g/cm^3) is determined by the paraffin-clod method as described by Chleborad and others (1975). Oven-dry peds are thinly coated with paraffin and weighed both in air and submerged in water. Bulk-density measurements of replicate samples generally vary by less than 10 percent of the average value of the samples.

Advantages and Limitations of the cS Index

The cS index is a superior measurement of calcic soil development because it integrates three soil parameters: $CaCO_3$ content, thickness, and bulk density. The procedure for calculating cS values is relatively straight-forward, and the analyses are simple and inexpensive to perform with equipment available in most earth science laboratories. Calcic soils formed in materials ranging from gravels to clays can be compared by this technique, and if parent materials that have little or no primary carbonate are selected, then most of the carbonate present in the soil (cT) must be a secondary component (cS), thereby enhancing the accuracy of the cS determination. Because the primary carbonate content of calcic soils is estimated from parent materials and can be controlled by sampling design, calcic soils differ significantly from other soils whose developmental indices are based on primary constituents such as clay. For example, clay in desert soils could be an original constituent of the parent material, an eolian contribution, or a product of in situ weathering of mineral grains; in most cases these three clays are analytically inseparable. Layered parent materials, such as alluvium, often have depositional variations in clay content and commonly become progressively finer-grained upwards. Thus, the initial distribution of clay in some parent materials can be masked or enhanced by pedogenic clay in B horizons.

There are two potential limitations to the cS index, both of which can be controlled by careful sampling design. The first limitation is the difficulty of determining the bulk density of weak calcic soils formed in coarse materials, many of which lack peds (coherent blocks). This problem is circumvented by taking a large sample and by determining both its extraction weight and its volume. Alternatively, in situ bulk density can be determined using a nuclear-density probe, as is done in road construction to measure the emplacement density and moisture content of compacted aggregate. Because nuclear-density probes do not disturb the soil, replicate measurements can be made on the same part of the soil.

A second, more serious limitation arises for soils in calcareous parent materials. Much of these soils' cT content is a primary component. Redistribution of primary $CaCO_3$ or errors in determining cP will decrease the precision of the computed cS value. In terms of this technique, soils in coarse-grained, calcareous parent materials are the most difficult to analyze: they are best avoided in favor of more suitable materials.

RELATION BETWEEN CARBONATE MORPHOLOGY CA^{++} INFLUX, AND CLIMATE

The time required to form successive stages of carbonate morphology in different regions is related to the age of the soil, the texture of the soil, the rate of influx of solid $CaCO_3$ and soluble Ca^{++}, and the amount and distribution of annual rainfall. By comparing the maximum stage of carbonate morphology in calcic soils from eight chronosequences in the Southwest, I found that significant regional differences in the time-dependent formation of these stages result primarily from various combinations of climate and Ca^{++} and $CaCO_3$ influx.

The eight chronosequences consist of noncalcareous, gravelly alluvium that has constructional geomorphic surfaces bearing relict calcic soils (Table 2). The vertical placement of these geologic units and (or) geomorphic surfaces in Table 2 reflects age estimates based on published and unpublished data. Correlations of units in different regions, however, are based partly on the climatic model of depositional and erosional cycles used by Hawley and others (1976).

Table 2 shows the maximum stage of carbonate morphology found in relict soils of each chronosequence. One of the points to illustrate here is the variation in stages observed in soils of the same age over broad geographic transects. For example, Stage IV morphology does not occur in any of the soils of the upper Arkansas River valley (column 1), yet Stage IV morphology is found on progressively younger soils southward into New Mexico. The areas in Table 2 form two geographic transects: a north-to-south transect from central Colorado to southern New Mexico (a distance of 800 km) and an east-to-west transect from eastern New Mexico to southeastern California (a distance of 850 km). The mean-annual precipitation and temperature values (Table 2; U.S. National Oceanic and Atmospheric Administration, 1978) are gross indicators of the soil-moisture regime in each of these areas and are included to aid in comparison of $CaCO_3$ morphologies. On the basis of their climate, the study areas are grouped in four broad categories: cool-arid, temperate-semiarid, warm-semiarid, and hot-arid.

As one many expect, this regional study shows that most calcic soils cannot be correlated solely on the basis of $CaCO_3$ morphology. Even within the same climatic grouping, soils of a single age can vary by a full stage of morphologic development over large regions. For example, calcic soils in 200,000- to 400,000-year-old alluvium along the Rio Grande in New Mexico (Table 2, columns 3, 4, and 5) have a maximum Stage III morphology in the northern and central part of the State, but a maximum Stage IV morphology in the southern part.

Correlations based on morphology alone might be misleading because some morphologies form over intervals of as much as hundreds of thousands of years. Near San Acacia, New Mexico (Fig. 1 and Table 2, column 4), soils in late to middle Pleistocene alluvium (units D, E, F, G, and H, and the Cliff surface) all have Stage III morphology, and although the older units have thicker soils, $CaCO_3$ morphology fails as a distinguishing criterion for these soils, which range from about 100,000 to about 500,000 years in age.

Calcic soils formed in limestone-rich alluvium have a plentiful source of primary calcium carbonate for in situ leaching and thus develop faster than soils on noncalcareous alluvium. The fastest morphologic development we have seen in the Southwest

TABLE 2. MAXIMUM STAGES OF CARBONATE MORPHOLOGY IN GRAVELLY RELICT SOILS DEVELOPED IN ALLUVIAL UNITS OR BELOW CONSTRUCTIONAL SURFACES OF UNNAMED ALLUVIAL UNITS IN THE SOUTHWESTERN UNITED STATES

[Soils that have the same stages of CaCO₃ morphology are shown by tie lines. Alluvial units designated by letters and those shown with "alluvium" lowercased are informally named; all other names of alluvial units are formal. Names of geomorphic surfaces shown in italics. Ages are shown on a nonlinear scale. Climatic data from U.S. National Oceanic and Atmospheric Administration (1978).]

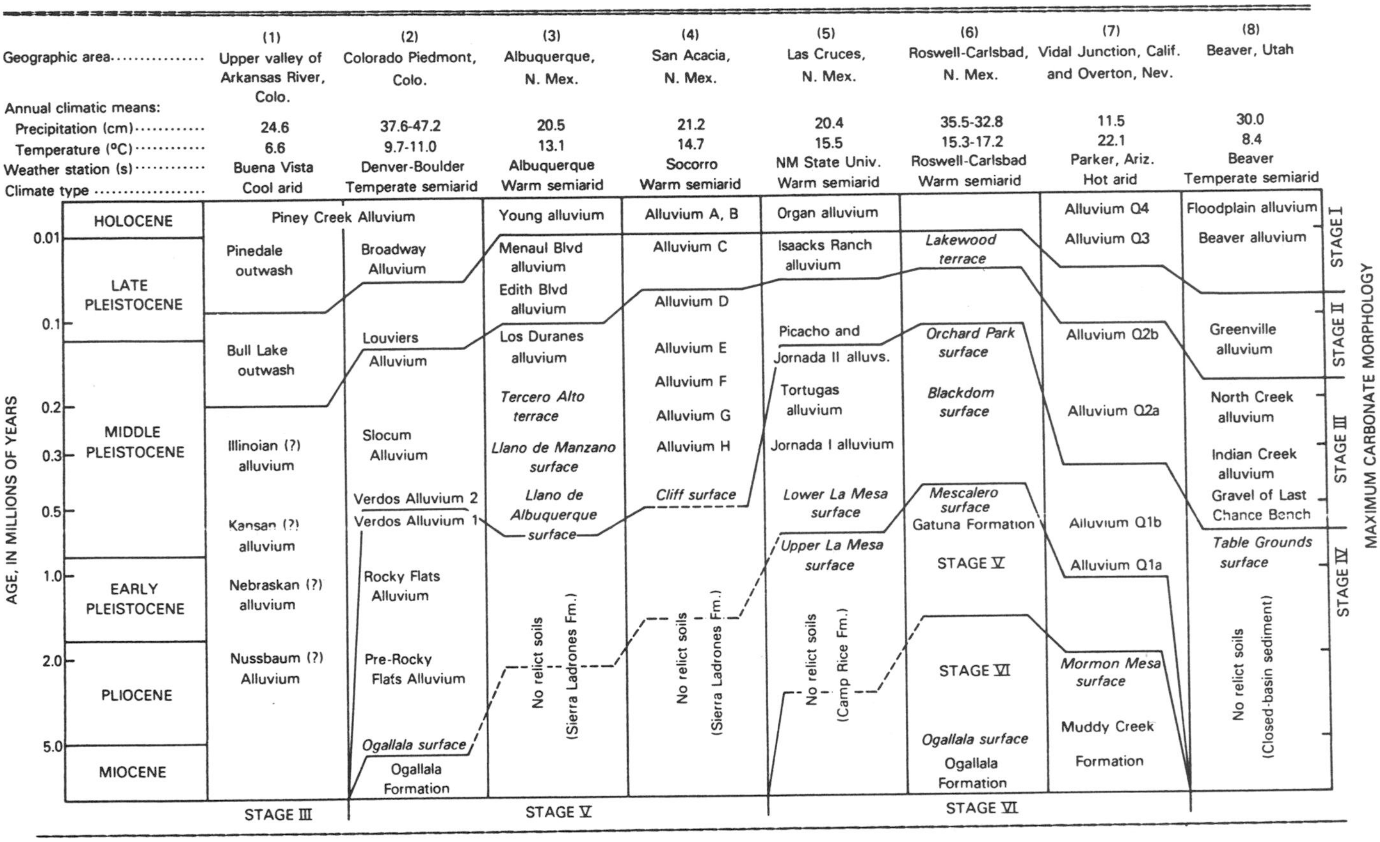

	(1)	(2)	(3)	(4)	(5)	(6)	(7)	(8)
Geographic area	Upper valley of Arkansas River, Colo.	Colorado Piedmont, Colo.	Albuquerque, N. Mex.	San Acacia, N. Mex.	Las Cruces, N. Mex.	Roswell-Carlsbad, N. Mex.	Vidal Junction, Calif. and Overton, Nev.	Beaver, Utah
Annual climatic means:								
Precipitation (cm)	24.6	37.6-47.2	20.5	21.2	20.4	35.5-32.8	11.5	30.0
Temperature (°C)	6.6	9.7-11.0	13.1	14.7	15.5	15.3-17.2	22.1	8.4
Weather station(s)	Buena Vista	Denver-Boulder	Albuquerque	Socorro	NM State Univ.	Roswell-Carlsbad	Parker, Ariz.	Beaver
Climate type	Cool arid	Temperate semiarid	Warm semiarid	Warm semiarid	Warm semiarid	Warm semiarid	Hot arid	Temperate semiarid

Age, in millions of years (nonlinear scale): 0.01 — HOLOCENE; 0.1 — LATE PLEISTOCENE; 0.2, 0.3, 0.5 — MIDDLE PLEISTOCENE; 1.0, 2.0 — EARLY PLEISTOCENE; 2.0 — PLIOCENE; 5.0 — MIOCENE.

Epoch / age	(1)	(2)	(3)	(4)	(5)	(6)	(7)	(8)	Max. carbonate morphology
HOLOCENE	Piney Creek Alluvium	Broadway Alluvium	Young alluvium	Alluvium A, B	Organ alluvium	*Lakewood terrace*	Alluvium Q4 / Alluvium Q3	Floodplain alluvium	STAGE I
LATE PLEISTOCENE	Pinedale outwash		Menaul Blvd alluvium; Edith Blvd alluvium	Alluvium C; Alluvium D	Isaacks Ranch alluvium			Beaver alluvium	STAGE II
	Bull Lake outwash	Louviers Alluvium	Los Duranes alluvium	Alluvium E; Alluvium F	Picacho and Jornada II alluvs.	*Orchard Park surface*	Alluvium Q2b	Greenville alluvium	
MIDDLE PLEISTOCENE	Illinoian (?) alluvium	Slocum Alluvium	*Tercero Alto terrace*; *Llano de Manzano surface*	Alluvium G; Alluvium H	Tortugas alluvium; Jornada I alluvium	*Blackdom surface*	Alluvium Q2a	North Creek alluvium; Indian Creek alluvium	STAGE III
	Kansan (?) alluvium	Verdos Alluvium 2; Verdos Alluvium 1	*Llano de Albuquerque surface*	*Cliff surface*	*Lower La Mesa surface*; *Upper La Mesa surface*	*Mescalero surface* Gatuna Formation	Alluvium Q1b	Gravel of Last Chance Bench; *Table Grounds surface*	STAGE IV
EARLY PLEISTOCENE	Nebraskan (?) alluvium	Rocky Flats Alluvium	No relict soils (Sierra Ladrones Fm.)	No relict soils (Sierra Ladrones Fm.)	STAGE V; No relict soils (Camp Rice Fm.)	STAGE V	Alluvium Q1a	No relict soils (Closed-basin sediment)	
PLIOCENE	Nussbaum (?) Alluvium	Pre-Rocky Flats Alluvium				STAGE VI	*Mormon Mesa surface*; Muddy Creek Formation		
MIOCENE		*Ogallala surface* Ogallala Formation				*Ogallala surface* Ogallala Formation			

Bottom stage labels: STAGE III (col 1); STAGE V (cols 2–5); STAGE VI (cols 6–8).

References for geologic and soils data used in the above columns: (1) Scott (1975); G.R. Scott and R.R. Shroba, written commun., 1977; (2) Modified from Scott (1963) and Machette and others (1976); (3) Modified from Lambert (1968), Bachman and Machette (1977). All alluvial units are informally named. (4) Bachman and Machette (1977), modified from Machette (1978a); (5) Modified from Gile and others (1979 and 1981), Machette, unpubl. data. All alluvial units are informally named. (6) Modified from Bachman (1976) and Hawley and others (1976); (7) Modified from Gardner (1972), Bull (1974), and Ku and others (1979); Machette, unpubl. data; and (8) Machette (1982); Machette, unpubl. data. All alluvial units are informally named.

is in the Roswell-Carlsbad area of southeastern New Mexico, where relict soils on the three youngest surfaces, the Lakewood, Orchard Park, and Blackdom (Table 2, column 6), are underlain by alluvium derived from limestone-rich terrain in the Sacramento Mountains. Additionally, the Roswell-Carlsbad area has both relatively abundant rainfall (33 to 35 cm/year) and an extensive, upwind source area of calcareous rocks which provide airborne Ca⁺⁺. The southeastern New Mexico area has an optimum combination of climate, Ca⁺⁺ influx, and limestone parent materials for in situ weathering.

Stages of CaCO₃ morphology are useful, though, for distinguishing and correlating soils and their associated geomorphic surfaces within local areas, such as individual drainage basins. CaCO₃ morphology can be used to differentiate some relict soils of less than 150,000 years age. In the Las Cruces region, soils in the Organ, Isaacks Ranch, Jornada II, and Picacho alluvial units (all informal terms) have diagnostic stages of morphology (Table 2, column 5). The latter units, the Jornada II and Picacho, are the same age and contain soils that have a maximum morphology of Stage III (in fine-grained material) to weak Stage IV (in gravel). The soils in the next significantly younger unit, the latest Pleistocene Isaacks Ranch, contain carbonate nodules (Stage II) in fine-grained materials, whereas soils in similarly textured Holocene Organ alluvium only have filaments of carbonate (Stage I). Thus, these groups of alluvium can be distinguished in the field on the basis of carbonate morphology. Many of the soils in the older alluvial units, such as the Jornada I, Tortugas, and parts of the Picacho, all have a maximum Stage IV morphology and are not so easily differentiated.

MODEL OF CaCO₃ ACCUMULATION

The maximum carbonate morphology in calcic soils (Table 2) observed along two geographic transects of the Southwest results from differences in their respective rates of CaCO₃ accumulation, which are controlled by a delicate balance between the supply of Ca^{++} ions and the amount and effectiveness of rainfall in moving Ca^{++} into the soil. Junge and Werby's (1958, Fig. 7) analyses show Ca^{++} is plentiful in rainwater of the Southwest, particularly in the Four Corners region. The rainfall component of Ca^{++} provides a broad regional base for calcic soil formation and this component can be a major part of the total Ca^{++} influx in many areas where sources of solid CaCO₃ are limited or not present.

Variations in the total Ca^{++} influx both between and within regions are also caused by locally derived solid carbonate. For example, there is a moderate difference between the average rate of Ca^{++} accumulation for central and southern New Mexico, yet these areas have similar amounts of rainfall and dissolved Ca^{++}. Therefore, the differences in rates of accumulation must be due to local sources of solid carbonate influx, such as the large areas of slightly calcareous eolian sand in the southern part of the state.

Rainfall and Ca⁺⁺ Influx Control of Potential Rates

In the Southwest, the potential rate of carbonate accumulation appears to be controlled mainly by the supply of Ca^{++} to the soil surface and the amount of moisture (rainfall) available to move Ca^{++} in the soil. Because soil carbonate can form under greatly varying conditions of rainfall and Ca^{++} influx, the rates also vary widely under semiarid and arid climates of the Southwest. To illustrate these conditions, I have plotted rates of carbonate accumulation for soils of the eight chronosequences against a parameter that incorporates their relative amounts of moisture, largely as measured by annual rainfall ("moisture" shown on the y-axis of Fig. 5), and against their relative influxes of Ca^{++} (combined solid CaCO₃ and Ca^{++} in rainwater). The influx scale is open ended and ranges from no influx to high influx. Points on this diagram are plotted from long term accumulation rates (see also Table 2), partly on informed estimates of local and regional solid carbonate influx rates, and on the dissolved-Ca^{++} data of Junge and Werby (1958).

Leaching, however, is greatly dependent on the supply of Ca^{++}. Leaching can occur under moderate rainfall if the supply of Ca^{++} is low, but requires greater amounts of rainfall if the supply

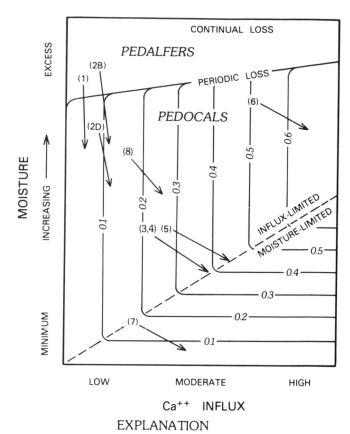

EXPLANATION

1. *Buena Vista, CO*	5. *Las Cruces, NM*
2. *Boulder-Denver, CO*	6. *Roswell-Carlsbad, NM*
3. *Albuquerque, NM*	7. *Vidal Junction, CA*
4. *San Acacia, NM*	8. *Beaver, UT*

Figure 5. Carbonate accumulation rates (R̄x, shown by vertical and horizontal lines, in g of CaCO₃/cm²/1,000 years) under varying conditions of moisture and Ca^{++} influx. Soil data for the eight chronosequences (Table 2) are plotted under the inferred conditions and rates of accumulation during pluvial episodes; arrows show inferred conditions during interpluvial episodes. Most areas have higher accumulation rates during interpluvials; Vidal Junction (7), an exception, has slower accumulation rates during interpluvials.

of Ca^{++} is high. The upper limit of moisture under which pedocals form is shown by an inclined boundary in Figure 5.

The field of carbonate accumulation is divided into two subfields: moisture-limited and influx-limited (Fig. 5). The line separating these subfields represents equilibrium conditions at which there is a balance between Ca^{++} supply and moisture. When there is excess Ca^{++}, that is, Ca^{++} that is not being fully leached from the surface of the soil, the potential rate of carbonate accumulation is not realized, and the conditions are termed "moisture-limited." An increase in the limited factor, or a balanced increase in both factors, will result in a net increase in the rate of CaCO₃ accumulation.

Moisture-limited areas have a greater Ca^{++} influx than can be accommodated by local rainfall and soil-moisture conditions, hence the potential rate of carbonate accumulation is limited because all the supplied Ca^{++} cannot be translocated into the soil. In these cases, carbonate accumulation rates are increased by increasing moisture, but not the Ca^{++} influx, because there is already an overabundance of Ca^{++}. Soils from such areas typically have silt-rich calcareous A horizons that may or may not overlie noncalcareous B horizons. Soils of the Vidal Junction area (eastern Mojave Desert) and the Mormon Mesa area (southern Nevada, Fig. 1) are probably moisture-limited at present.

Conversely, influx-limited areas have a greater amount of moisture available for translocation of Ca^{++} than is supplied to the soil surface by rainfall and solid debris. Such areas, especially those where rapid snowmelt occurs each spring, can have excess leaching potential with respect to Ca^{++} influx, particularly if rates of Ca^{++} influx are low. The upper 1 to 2 m of soils in these areas are typically leached of carbonate. The field under which these conditions exist bounds the lower limit of the pedalfers (Fig. 5). Although the contours of equal rate parallel either the x- or y-axis in Figure 5, there may be an inherent dependence in the amount of rainfall and amount of dissolved Ca^{++} (Junge and Werby, 1958). Influx-limited conditions probably exist in areas such as the valley of the upper Arkansas River near Buena Vista, Colorado (Table 2, column 1), and the Colorado Piedmont between Denver and Boulder, Colorado (Table 2, column 2). These conditions also exist near the pedocal-pedalfer boundary throughout the Southwest, especially in mountainous areas, and are evidenced by laterally discontinuous accumulations of soil carbonate. Relict pedalfers and pedocals coexist in the middle Pleistocene Verdos Alluvium along the pedocal-pedalfer boundary northwest of Denver, Colorado (Machette and others, 1976). These pedocals are the remnants of extensive, continuous calcic soils that accumulated carbonate when the boundary was at higher elevations in the past.

On the basis of soils data (Gile, 1975, 1977; Machette, unpublished data, 1984) and regional studies of paleoecology (Spaulding and others, 1983) it seems plausible that semiarid areas in central and southern New Mexico have been more arid during interpluvial episodes (such as the Holocene) than during pluvial episodes, resulting in less vegetative cover and a greater influx of airborne Ca^{++} and $CaCO_3$. Although the potential for carbonate accumulation would have increased during the interpluvials because of higher influx of carbonate, the increase probably was partly offset by a decrease in rainfall. The New Mexico soils (Fig. 5; areas 3, 4, 5, and 6) are plotted on the basis of their average rates of carbonate accumulation ($R\bar{x}$) which, in these areas, are probably more indicative of pluvial conditions than interpluvial conditions as shown below. If this model based on Ca^{++} influx and moisture is correct, soils forming in these four regions would plot further down and to the right in the moisture-limited field during interpluvials (times of increased moisture but decreased Ca^{++} influx). Their plotted positions, which reflect conditions during pluvials, lie within the influx-limited field.

In other parts of the Southwest, the effects of climatic change on landscape stability and carbonate accumulation rates could be much different from those postulated for New Mexico. For example, the rainfall over much of the Southwest is less than 20 cm annually, and in some areas it is less than 10 cm. Calcic soils in moisture-limited areas, such as Vidal Junction (area 7 on Fig. 5 and Table 2), may have formed more slowly in the Holocene than in the late Pleistocene. Calcic soils in arid regions commonly have an excess supply of Ca^{++} relative to their limited amount of rainfall. Therefore, a substantial increase in the rainfall of these areas (such as during the pluvials) would allow more Ca^{++} to be leached into the soils and thereby increase the rate of carbonate accumulation.

Analyses of C^{14}-dated pack-rat middens from the Southwest suggest that vegetation zones during latest Pleistocene time were significantly lower than at the present (Van Devender and Spaulding, 1979; Spaulding and others, 1983) probably because a substantially wetter and (or) cooler climate prevailed during the late Pleistocene. The Vidal Junction area in the eastern Mojave Desert presently receives about 15 cm of rainfall, the majority of which falls during high-intensity summer storms. A modest increase of 15 cm (to a total of 30 cm) would double the amount of annual rainfall and probably would be accompanied by a marked increase in the amount of vegetational cover, to the extent that the potential sources of Ca^{++} would be greatly reduced. A climatic change of this magnitude could cause soil conditions to transgress the equilibrium line, such that there are influx-limited conditions during pluvials and moisture-limited conditions during interpluvials.

A condition not yet discussed is one of high accumulation under high Ca^{++} influx and moderate amounts of rainfall (annual mean of about 34 cm). The long-term rates of accumulation for soils of the Roswell-Carlsbad area (Fig. 5, area 6) indicate such conditions. Values for this area probably plot within the influx-limited field now (interpluvial), but during pluvials, the soil-moisture conditions may have approached those of the pedalfers (Fig. 5, upper part). Soil conditions in the Boulder-Denver area and in the upper Arkansas River Valley (Table 2) probably are analogous to those of the Roswell-Carlsbad area, except that they have a much lower supply of Ca^{++}.

RATES OF CALCIUM CARBONATE ACCUMULATION

Soil carbonate accumulation rates vary over time and between localities in response to geographic, geologic, and climatic controls. To make accurate soil-age estimates from cS data, one must understand the magnitude and frequency of these possible variations. These variations are detected by comparing the cS content of soils that have accumulated carbonate during several known time intervals (such as the past 10,000, 50,000, 100,000, and 500,000 years), thereby yielding rates of accumulation.

Rates of carbonate accumulation are influenced by changes in climate both in time and over regions. I used Gile and others'

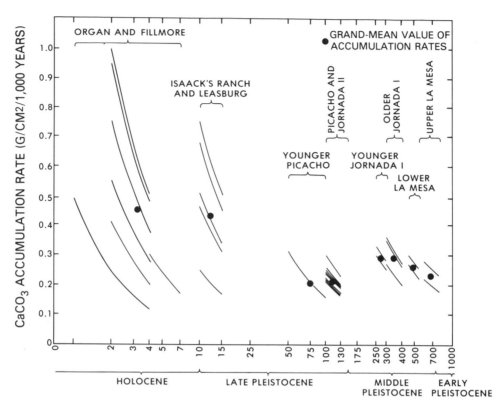

Figure 6. Average carbonate-accumulation rates ($R\bar{x}$) of soils formed in Pleistocene and Holocene alluvial units of the Las Cruces area, New Mexico. Alluvial units are those used by Gile and others (1979) and Gile and others (1981, Table 25) to calculate cT contents. Solid circles indicate the grand-mean values of possible average rates for soils in each alluvial unit.

(1981, Table 25) cT data to calculate rates of carbonate accumulation in the Las Cruces region. Data were selected for soils in noncalcareous parent materials, so that their cP component would be small. Their cT values would therefore approximate cS. However, the Las Cruces soils were not sampled specifically for analysis of carbonate. The older soils generally had thick sampling intervals in which there could be large but undetected variations in carbonate content or bulk density. Also, some of these soils are eroded and others were not sampled deeply enough to penetrate noncalcareous material (Cn horizon). Therefore, some of the Las Cruces soils may have cT values that are minimum estimates of cS, not maximum estimates.

Because the ages of some of the Las Cruces soils are not closely defined, I use a "probable age range" in calculating the average rates of carbonate accumulation. For example, the bulk of the Organ alluvial unit is considered to be between 1,000 and 4,000 years old, based on radiocarbon dates from organic carbon in the alluvium. To illustrate the effect that age ranges have on the calculation of accumulation rates, assume that a soil in this unit had 1.25 g of $CaCO_3$ per square-centimeter column. The carbonate accumulation rate for this soil is 0.125 to 0.50 $g/cm^2/1,000$ years (cS divided by age range). Such values represent the average rate ($R\bar{x}$) of carbonate accumulation

because the soils are relicts; that is, they are the cumulative products of continuous soil formation. The ages used are those I consider geologically reasonable based on degree of soil development, C^{14} dates on inorganic and organic carbon, correlations of the Las Cruces chronosequence with other dated alluvial units in New Mexico, geomorphic considerations, and unpublished uranium-trend soil ages determined by J. N. Rosholt (written commun., 1981). Ages for the Las Cruces soils and alluvium are modified from those of Gile and others (1981).

Carbonate Accumulation Rates Through Time

One way to analyze the magnitude of change of carbonate accumulation rates is to compare Holocene soils with their next older counterparts, late Pleistocene soils, thereby contrasting soils formed during different parts of the most recent climatic cycle. The Las Cruces soils are ideal for such a comparison. The possible ranges in their ages against their potential accumulation rates are shown on Figure 6 as a series of lines whose upper end points represent maximum rates and minimum ages and whose lower end points show the opposite. The actual rate of accumulation for any of these soils should lie somewhere on the resultant line (these lines are curves because of the logarithmic scale used on

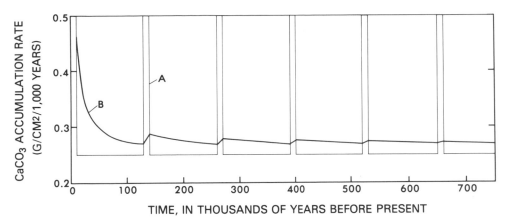

Figure 7. Model of calcium carbonate accumulation rates versus time. Line A shows the instantaneous rate (Ri) of $CaCO_3$ accumulation at a point in time. Line B shows average rate ($R\bar{x}$) through time. Model is based on interpluvial periods simulated by 10,000-year-long intervals in which Ri = 0.5 $g/cm^2/1,000$ years and pluvial periods simulated by 120,000-year-long intervals in which Ri = 0.25 $g/cm^2/1,000$ years.

the x-axis of Fig. 6). Figure 6 shows that the soils data plot as discrete groups of curves, and these groups range from Holocene and latest Pleistocene to middle Pleistocene age.

The young soils (Holocene and latest Pleistocene) have a wide range of possible accumulation rates for two reasons. First, they are weakly developed, and their cS contents are only a small part of their cT contents; hence, errors in determining their cP components greatly affect cS values. Secondly, the range of possible ages for each soil is large in comparison with the actual age. Conversely, the older groups of soils have smaller ranges of possible accumulation rates for exactly the opposite reasons.

Because of the range in values within any one group of soils, I computed the grand mean of $R\bar{x}$ values from the average value of each soil; the grand means are shown by large dots on Figure 6. The grand means clearly show that Holocene and latest Pleistocene calcic soils have average accumulation rates that are higher than those of older soils. The grand means for soils of Holocene and latest Pleistocene age are 0.46 and 0.43 g of $CaCO_3/cm^2/1,000$ years, respectively. These rates represent the accumulation of carbonate over intervals of less than 5,000 to as much as 18,000 years (the older limit for latest Pleistocene alluvium). Nevertheless, these rates largely reflect soil formation under Holocene interpluvial conditions (the past 10,000 years).

Soils in the Picacho and Jornada II alluvial units, the majority of which are between 75,000 and 150,000 years old, were formed mostly under pluvial conditions of the last complete climatic cycle, whereas older soils were formed under one or more complete climatic cycles. These soils, ranging back to middle or early Pleistocene age, have grand-mean accumulation rates of 0.21 to 0.29 g of $CaCO_3/cm^2/1,000$ years; values of about one-half of the younger rates. By using only grand means of accumulation rates, the effect of anomalous rates that could result from mistaken age calls or from errant determinations of cS should be minimized.

These data clearly show that soils in southern New Mexico have accumulated carbonate under conditions of an interpluvial climate at an average rate that is nearly twice that which prevailed during the preceding pluvial climatic episode. If one assumes that these rate changes are climatically controlled, then the average accumulation rate through time can be analyzed with an algebraic step function. To simply illustrate such a step function (Fig. 7), the interpluvial periods of southern New Mexico are simulated by 10,000-year-long intervals of high carbonate accumulation (0.5 $g/cm^2/1,000$ years) and the pluvial periods are simulated by 120,000-year-long intervals of low accumulation (0.25 $g/cm^2/1,000$ years). This model, although simpler in detail than most pluvial-interpluvial cycles currently being considered, calls for accumulation rates in southern New Mexico to double during interpluvials, an assumption I consider reasonable based on soil data (Fig. 6).

The instantaneous rate of carbonate accumulation (Ri) is herein defined as the accumulation rate at any point in time. In the step-function model (Fig. 7), the instantaneous rate changes between 0.25 and 0.50 $g/cm^2/1,000$ years as shown by line A. From these instantaneous values, I calculated the average rate of accumulation ($R\bar{x}$) for progressively longer intervals of time before the present (Fig. 7, line B). Soils that formed during one complete climatic cycle (130,000 years) or longer have similar $R\bar{x}$ values because fluctuations caused by short, high-rate interpluvials have little impact on the overall rate established during the long pluvial intervals.

A model based on these instantaneous rates shows that long intervals of low accumulation during pluvial episodes have a pronounced dampening effect on the short intervals of high accumulation during the interpluvials. Although one might disagree with the details of this model, any model that incorporates rate oscillations with long and short intervals of this amplitude will produce the same basic trend in carbonate accumulation rates. For example, increasing the number of short, high-rate intervals within the last climatic cycle superimposes minor fluctuations on

TABLE 3. AGE, MAXIMUM STAGE OF CARBONATE MORPHOLOGY, SECONDARY CARBONATE
CONTENT (cS), AND AVERAGE RATE OF SECONDARY CALCIUM CARBONATE
ACCUMULATION (R$\bar{\text{x}}$) IN CALCIC SOILS OF FOUR REGIONS
OF THE AMERICAN SOUTHWEST

Area: Geomorphic surface	Soil age (m.y.)	Stage of CaCO$_3$	cS, in g/cm^2 Mean	Range	R$\bar{\text{x}}$, in g/cm^2 per 1,000 years
Beaver, Utah: Last Chance Bench	0.5	III+	71	64-78	0.14$\pm$0.01
Albuquerque, New Mexico: Llano de Albuquerque	0.5	III	110	98-114	0.22$\pm$0.02
Las Cruces, New Mexico: Lower La Mesa	0.5	IV	129	120-137	0.26$\pm$0.02
Roswell-Carlsbad, New Mexico: Mescalero	0.5	V-	257	229-307	0.51$\pm$0.06

Data for soils near Beaver, Utah, are from Machette (1982) and Machette, unpublished data. Most of the data for the Las Cruces chronosequence are from Gile and others (1979). The remaining New Mexico data are from Bachman and Machette (1977) and Machette, unpublished data.

the basic curve and results in a slightly higher average rate of accumulation.

If the basic structure and assumptions of this model are valid, then soils that formed for more than about 100,000 years should have similar average accumulation rates and cS values that can be used to correlate soils locally and to estimate the ages of calcic soils older than about 100,000 years. The ages of soils younger than 100,000 years can also be estimated, although such estimates may be off because the average accumulation rate during this short-term interval and those used for the long-term (500,000-year) interval may be significantly different.

Regional Variations in Carbonate Accumulation Rates

The amount of regional variation in average CaCO$_3$ accumulation rates was determined by comparing cS values of relict calcic soils formed on correlative middle Pleistocene sediments of three areas in New Mexico. These values were also compared with one obtained from an area in central Utah, 600 to 1,200 km to the northwest, to determine their degree of similarity.

Relict middle Pleistocene calcic soils about 500,000 years old (Hawley and others, 1976; Bachman and Machette, 1977) are preserved along the Rio Grande between Albuquerque and Las Cruces, New Mexico (Fig. 1), just below the constructional surfaces of the Sierra Ladrones Formation and the Camp Rice Formation of Strain, 1966, and in correlative piedmont-slope sediments that rise gently mountainward from the Rio Grande. Likewise, east of the Pecos River in southeastern New Mexico, the "Mescalero caliche" (a pedogenic calcrete) is developed in the Gatuna Formation (Bachman, 1976). The Gatuna locally contains water-laid beds of the Lava Creek ash, a 600,000-year-old volcanic ash erupted from calderas in the Yellowstone area of northwestern Wyoming (Izett and Wilcox, 1982). Because the

ash is interbedded with, rather than overlying, the Gatuna Formation, the Mescalero probably began to form immediately after deposition of the Gatuna Formation, about 500,000 years ago.

Calcic soils of the same age are present in unconsolidated alluvium near Beaver, Utah (Fig. 1), which is in an intermontane basin on the boundary between the Basin and Range and the Colorado Plateaus provinces. The basal part of the gravels of Last Chance Bench, a thin but widespread piedmont-slope deposit, is interbedded with a 530,000-year-old rhyolitic pumice (Machette, 1982). Antiformal uplift coupled with rapid lowering of base level within the basin caused dissection of the gravel soon after it was deposited, thereby preventing further deposition on the gravels after 500,000 years ago. A strong Stage III calcic soil has formed in the gravels of Last Chance Bench during the past 500,000 years.

The cS contents of 500,000-year-old calcic soils in these four areas are shown in Table 3. The soils near Beaver are the least calcareous; they have an average cS of 71 g/cm^2 and an average accumulation rate (R$\bar{\text{x}}$) of 0.14$\pm$0.01 g/cm^2/1,000 years. Near Albuquerque and Las Cruces, New Mexico, soils of this age have cS contents of 110 and 129 g/cm^2 and R$\bar{\text{x}}$ values of 0.22 and 0.26 g/cm^2/1,000 years, respectively. Along the Rio Grande, the 500,000-year-old soils show a slight southward increase in the maximum stage of carbonate morphology from Stage III+ to IV (Table 3). This increase must be primarily the result of a slightly higher solid-CaCO$_3$ influx in southern New Mexico than in central or northern New Mexico, inasmuch as the modern temperature and rainfall values (Table 2) and amounts of Ca^{++} dissolved in rainfall (Junge and Werby, 1958) are similar along the Rio Grande from Albuquerque to Las Cruces.

In the Roswell-Carlsbad area, the Mescalero has weak Stage V development (Tables 2 and 3) and thus, is more advanced than soils of the same age elsewhere in New Mexico. Both the amount of rainfall and the rate of Ca^{++} influx in southeastern New Mexico are higher than in central and southern New Mexico, and these conditions result in a high rate of carbonate accumulation. The Mescalero has an average cS content of 257 g of $CaCO_3/cm^2$; about 2.5 times that of soils of the Llano de Albuquerque. The Mescalero has accumulated carbonate at an average rate of 0.51 ± 0.06 g/cm^2/1,000 years for the past 500,000 years; this is the highest rate yet determined in the Southwest.

GEOLOGIC APPLICATIONS OF CARBONATE DATA

The stratigraphy and geomorphology of alluvial terraces and adjacent piedmont-slope sediments along the Rio Grande in New Mexico were recently correlated by Hawley and others (1976). To illustrate the potential that cS data have in such correlations, I suggest correlations for some of these same sediments based on new age control from uranium-trend ages by the method of Rosholt (1980), on K-Ar dates obtained from associated volcanic rocks, and on soil-age estimates. Soil ages are estimated from cS data and the long-term accumulation rates determined in each area (Table 3). For example, if a soil has a cS content of 25 g/cm^2 and an average accumulation rate ($R\bar{x}$) of 0.20 g/cm^2/1,000 years, the estimated soil age is 125,000 years. Estimated soil ages of less than about 50,000 years are probably maximum estimates, because their average accumulation rates were probably higher than the long-term rate used in the step-function model (Fig. 7).

Quaternary Geology along the Rio Grande

The Quaternary units of the Albuquerque area shown in Figure 8 are slightly modified from those established by Lambert (1968), but the conclusions concerning soil ages and correlations are my own. Lambert recognizes three major cut-and-fill alluvial units that form constructional surfaces: the alluviums of Menaul Boulevard (an unnamed surface), Edith Boulevard (the Primero Alto terrace), and Los Duranes (the Segundo Alto terrace). Bachman and Machette (1977) recognized alluvium that forms an additional terrace, the Tercero Alto, which is overlain by 190,000-year-old basalt flows of the Albuquerque Volcanoes (Bachman and others, 1975). An even higher, unnamed alluvium forms a widespread piedmont-slope surface, referred to either as the Sunport surface (Lambert, 1968) or the Llano de Manzano surface (Bachman and Machette, 1977), south of Albuquerque in the eastern part of the Albuquerque-Belen basin. The Llano de Manzano surface is graded to an alluvial terrace which lies 92 to 113 m above the modern Rio Grande; it is considerably lower than the Llano de Albuquerque surface (215 to 110 m) in the same basin. The Llano de Albuquerque, for the most part, is the upper constructional surface of the Sierra Ladrones Formation (Pliocene to middle Pleistocene). Although parts of the surface may have been isolated from further deposition in the early Pleistocene in response to local uplift, the majority of the surface is

considered to have become geomorphically stable about 500,000 years ago in middle Pleistocene time (Hawley and others, 1976; Bachman and Machette, 1977).

At San Acacia, near the southern end of the Albuquerque-Belen basin, the Rio Grande is joined from the west by the Rio Salado, a major tributary stream that is flanked by middle to upper Pleistocene alluvial terrace units and piedmont-slope units (Fig. 8 and Machette, 1978a). The Cliff surface, a local fault-controlled erosion surface cut on the Sierra Ladrones Formation (Machette, 1978c), is the oldest geomorphic surface in this area. On the basis of the Cliff surface's elevation and its soil's cS content, Machette (1978c) considers it to be slightly younger than the Llano de Albuquerque surface, whose nearest outcrop is about 20 km north of San Acacia.

In the Las Cruces area, the alluvial stratigraphy of Hawley and Kottlowski (1969) and the soil-geomorphology studies of Gile, Peterson, and Grossman (1979) and Gile and others (1981) provide a detailed framework (Fig. 8) to which I correlate units in the San Acacia and Albuquerque areas. The upper La Mesa surface, one of the oldest geomorphic surfaces near Las Cruces, was isolated from the depositional plain of the Rio Grande by uplift along the Robledo fault in the middle(?) or early(?) Pleistocene. The next younger surface, lower La Mesa, is the widespread upper constructional surface of the Camp Rice Formation of Strain (1966) and is correlative with the Llano de Albuquerque surface to the north. The upper part of the Camp Rice Formation locally contains the 600,000-year-old Lava Creek ash bed (Izett and Wilcox, 1982), thus the lower La Mesa surface and the youngest part of the Camp Rice Formation must be slightly less than 600,000 years old. The lower La Mesa surface is here considered to be about 500,000 years old on the basis of depositional and geomorphic considerations. In the piedmont areas adjacent to but mountainward of the Rio Grande, the Jornada I alluvial unit is the youngest constructional part of the Camp Rice Formation; it was still being deposited after the Rio Grande started downcutting about 500,000 years ago. The Tortugas, Jornada II, and Picacho alluvial units record successively lower levels of downcutting along the Rio Grande and tributary drainages during middle and late Pleistocene time. The youngest alluvial units in the Las Cruces area include the Issacks Ranch and Leasburg (latest Pleistocene age) and the Organ and Fillmore (Holocene age).

Correlation and Age Estimates

The correlation of alluvial units along the Rio Grande is greatly enhanced if they are based on soil ages estimated from both secondary $CaCO_3$ contents and accumulation rates and more traditional criteria (Fig. 8). For example, the alluviums that form the Llano de Manzano and Jornada I surfaces and alluvial unit H (Fig. 8) appear to be correlative on the basis of their stratigraphic and topographic positions. Calcic soils below the Llano de Manzano surface (Fig. 8) have a cS content of about 70 g/cm^2, and yield soil ages of about 320,000 years, assuming that

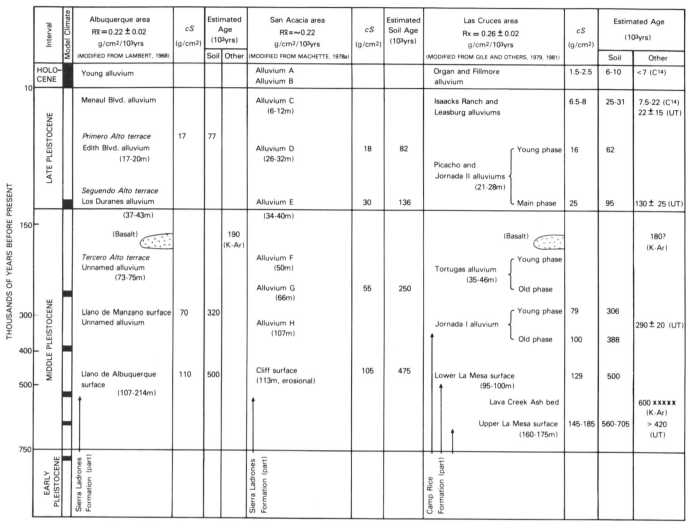

Figure 8. Tentative correlation of Quaternary alluvial units and constructional geomorphic surfaces (the latter shown in italics) along the Rio Grande in New Mexico. All of the alluvial units shown on here are informally named. Correlations are based on soil ages estimated from cS or cT contents, isotopic age data, relative position of unit in stratigraphic successions, and heights of constructional surface of units in meters above stream level (in parentheses) where applicable. $R\bar{x}$ is the average rate of carbonate accumulation (g of $CaCO_3/cm^2/1,000$ years) over the past 500,000 years. Time scale is nonlinear. Dark areas in "Model climate" column represent interpluvials; pluvials are white. This column is not intended to show the actual climatic chronology of the region, but reflects the simulation used in Figure 7.

$CaCO_3$ accumulated at an average rate of 0.22 g/cm²/1,000 years ($R\bar{x}$ for Albuquerque). In the San Acacia area, I was unable to determine the cS content of the soil in alluvial unit H, but the soil is stronger than that on alluvial unit G (estimated to be 220,000 years old) and weaker than that below the Cliff surface (estimated to be 475,000 years old). Thus, I use the midpoint between these age limits, about 350,000 years, for the soil in alluvial unit H. In the Las Cruces area, soils in a young phase of the Jornada I alluvium have an average cS content of 79 g of $CaCO_3/cm^2$, which yields an age of about 305,000 years. Thus, the cS data for these three soils indicate ages that range from 305,000 to 350,000 years, and undoubtedly would lie within concordant error limits if one could assign such values.

The age of the upper La Mesa surface is poorly constrained, but it is clearly older than the lower La Mesa surface (500,000 years), and sediments underlying both surfaces contain Pleistocene, but not Pliocene, vertebrate faunas (Hawley and others, 1976). The soils of the upper La Mesa surface have Stage V morphology, which is significantly more advanced than those of the lower La Mesa surface (Table 3). On the basis of cS contents of 145 to 185 g/cm², the soils of the upper La Mesa surface have been forming for 560,000 to 720,000 years. More importantly, the disproportionately advanced morphology of the soils of the upper La Mesa surface strongly suggests that they either have been partly eroded or have lost $CaCO_3$ through excess leaching. If cS contents from soils of the upper La Mesa surface (Fig. 8) are considered as minimum values, then the age of the tectonically uplifted upper La Mesa surface may be closer to one million years.

Other Applications

The above examples illustrate how the cS content of calcic soils can be used in making local and regional soil correlations and in estimating the ages of these soils when they are in Quaternary sediments. These estimates of soil age also provide information useful in a wide range of geologic studies, such as assessments of earthquake hazards, landform evolution, and paleoclimatic interpretation.

The recurrence intervals and the recency of fault movements that displace calcic soils can be estimated from the quantitative differences in soil development across fault zones. Near Albuquerque, calcic soils were displaced by four episodes of movement along the County Dump fault during the past 400,000 years. Machette (1978b) estimated the timing and amounts of displacement along this fault by relating the cS contents of the buried soils to their combined surface equivalent and found that the soil overlying the fault (soil U of Machette, 1978b) required 20,000 years to accumulate its carbonate. In retrospect, this estimate seems too old because the age of soil U was calculated from the long-term average rate of accumulation, whereas if Holocene rates are about 2 times faster as I now suggest, soil U could be about 10,000 years old. However, the age estimates for buried (faulted) soils (90,000 to 190,000 years each) are probably reasonable, because they formed during time intervals equal to or more than one complete climatic cycle.

SUMMARY

Calcic soils, and to a lesser extent pedogenic calcretes, are widespread in the semiarid and arid parts of the Southwest. Although these soils have formed in unconsolidated sediments as old as late(?) Miocene age and as young as Holocene age, most of them have formed since middle or late Pleistocene time. Calcic soils form mainly by subaerial precipitation of carbonate that derived from Ca^{++} dissolved in rainwater and that is leached from solid airborne carbonate. As carbonate accumulates in the soil, its morphology develops progressively through recognizable stages that are reflected in enriched carbonate content, increased bulk density, and thickening of calcic horizons.

I have described a morphologic sequence of calcic soils that consists of six stages, the first four of which were originally defined by Gile and others (1966). Pedogenic calcretes are advanced forms of indurated calcic soils that display evidence of the three latter stages of morphology. Stage IV is characterized by carbonate-rich laminae less than 1 cm thick. The next stage of morphology, Stage V, is marked by thicker laminae and incipient to thick pisolites. The most advanced stage morphology, Stage VI, includes multiple generations of brecciation, pisolite formation, and recementation. Stage VI calcic soils can resemble marine limestones in both percent $CaCO_3$ and bulk density.

The total-profile index of secondary carbonate (cS) integrates three soil parameters: $CaCO_3$ content, thickness, and bulk density. Values of cS are easily computed from soil-description data and from simple laboratory analyses of soil carbonate content and bulk density. Many of the problems inherent in other types of quantitative soil indices can be minimized or negated by carefully selecting sampling sites in calcic soil chronosequences.

Long-term carbonate-accumulation rates reflect regional variations in the amount of soil moisture available for leaching of carbonate and in the airborne supply of Ca^{++} and $CaCO_3$. Carbonate accumulation rates in the Roswell-Carlsbad area of southeastern New Mexico over the past 500,000 years exceed 0.5 $g/cm^2/1,000$ years and are the highest in the Southwest. At the other extreme, calcic soils in the Beaver, Utah, area have accumulated carbonate at rates of about 0.14 $g/cm^2/1,000$ years; this rate is 25 to 75 percent of equivalent-age soils in New Mexico.

Soil cS contents can be used to correlate and differentiate Quaternary sediments and, in some cases, can be used to estimate the ages of relict calcic soils and pedogenic calcretes. Studies of calcic soils are relevant to geologic problems such as the siting of critical facilities, the evaluation of earthquake and other natural hazards, the analysis of landform evolution, and the understanding of surficial geologic processes.

ACKNOWLEDGMENTS

The author is indebted to George Bachman for guidance and inspiration during this study. I greatly appreciate the quantitative aspect of pedology that Peter Birkeland has instilled in me and in many of his other students, and I have benefited greatly from numerous conversations with Leland Gile, John Hawley, and Fred Peterson, principal investigators of the U.S. Soil Conservation Service's study of desert soils. Leonard Gardner, Wally Hansen, Richard Hereford, Fred Peterson, and Gerald Richmond provided thoughtful criticism of the manuscript.

REFERENCES CITED

Aristarain, L. F., 1970, Chemical analyses of caliche profiles from the High Plains, New Mexico: Journal of Geology, v. 78, no. 2, p. 201–212.

Bachman, G. O., 1976, Cenozoic deposits of southeastern New Mexico and an outline of the history of evaporite dissolution: U.S. Geological Survey Journal of Research, v. 4, p. 135–149.

Bachman, G. O., and Machette, M. N., 1977, Calcic soils and calcretes in the southwestern United States: U.S. Geological Survey Open-File Report 77-794, 163 p.

Bachman, G. O., Marvin, R. F., Mehnert, H. H., and Merritt, V., 1975, K-Ar ages of the basalt flows at Los Lunas and Albuquerque, central New Mexico: Isochron/West, no. 13, p. 3–4.

Birkeland, P. W., 1974, Pedology, weathering, and geomorphological research: New York, Oxford University Press, 285 p.

—— 1984, Soils and Geomorphology: New York, Oxford University Press, 372 p.

Blake, W. P., 1902, The caliche of southern Arizona—An example of deposition by the vadose circulation: American Institute of Mining Engineers Transactions, v. 31, p. 220–226.

Bretz, J. H., and Horberg, L., 1949, Caliche of southeastern New Mexico: Journal of Geology, v. 57, p. 491–511.

Brown, C. N., 1956, The origin of caliche in the northeast Llano Estacado, Texas: Journal of Geology, v. 46, p. 1–15.

Bryan, K., and Albritton, C. C., Jr., 1943, Soil phenomena as evidence of climatic changes: American Journal of Science, v. 241, p. 469–490.

Bull, W. B., 1974, Geomorphic tectonic analysis of the Vidal region, *in* Information concerning site characteristics, Vidal Nuclear Generating Station [California]: Los Angeles, Southern California Edison Company, Appendix 2.5B, amendment 1, 66 p.

Chleborad, A. F., Powers, P. S., and Farrow, R. A., 1975, A technique for measuring bulk volume of rock materials: Association of Engineering Geologists Bulletin, v. 12, no. 4, p. 317–322.

Clarke, F. W., 1924, The data of geochemistry: U.S. Geological Survey Bulletin 770, 841 p.

Dreimanis, A., 1962, Quantitative determination of calcite and dolomite by using Chittick apparatus: Journal of Sedimentary Petrology, v. 32, no. 3, p. 520–529.

Elias, M. K., 1931, The geology of Wallace County, Kansas: Kansas Geological Survey Bulletin 18, 254 p.

Gardner, R. L., 1972, Origin of the Mormon Mesa caliche, Clark County, Nevada: Geological Society of America Bulletin, v. 83, no. 1, p. 143–156.

Gile, L. H., 1975, Holocene soils and soil-geomorphic relations in an arid region of southern New Mexico: Quaternary Research, v. 5, p. 321–360.

—— 1977, Holocene soils and soil-geomorphic relations in a semi-arid region of southern New Mexico: Quaternary Research, v. 7, p. 112–132.

Gile, L. H., and Hawley, J. W., 1966, Periodic sedimentation and soil formation on an alluvial-fan piedmont in southern New Mexico: Soil Science Society of America Proceedings, v. 30, no. 2, p. 261–268.

Gile, L. H., Hawley, J. W., and Grossman, R. B., 1981, Soils and geomorphology in the Basin and Range area of southern New Mexico—Guidebook to the Desert Project: New Mexico Bureau of Mines and Mineral Resources Memoir 39, 222 p.

Gile, L. H., Peterson, F. F., and Grossman, R. B., 1965, The K horizon—master soil horizon of carbonate accumulation: Soil Science, v. 99, no. 2, p. 74–82.

—— 1966, Morphological and genetic sequences of carbonate accumulation in desert soils: Soil Science, v. 101, p. 347–360.

—— 1979, The Desert Project soil monograph: Washington, U.S. Soil Conservation Service, 984 p.

Goudie, A., 1973, Duricrusts in tropical and subtropical landscapes: Oxford, Clarendon Press, 174 p.

Guthrie, R. L., and Witty, J. E., 1982, New designations for soil horizons and layers in the new *Soil Survey Manual*: Soil Science Society of America Journal, v. 46, p. 443–444.

Harden, J. W., 1982, A quantitative index of soil development from field descriptions: Examples from a chronosequence in central California: Geoderma, v. 28, no. 1, p. 1–28.

Harden, J. W., and Marchand, D. E., 1977, The soil chronosequence of the Merced River area [California], *in* Singer, M. J., ed., Soil development, geomorphology, and Cenozoic history of the northeastern San Joaquin Valley and adjacent areas, California: Guidebook for Joint Field Session, American Society for Agronomy, Soil Science Society of America, and Geological Society of America, Guidebook, Davis, University of California Press, p. 22–38.

Hawley, J. W., and Kottlowski, F. E., 1969, Quaternary geology of the south-central New Mexico border region, *in* Kottlowski, F. E., and LeMone, D. V., eds., Border stratigraphy symposium: New Mexico Bureau of Mines and Mineral Resources Circular 104, p. 89–104.

Hawley, J. W., Bachman, G. O., and Manley, K., 1976, Quaternary stratigraphy in the Basin and Range and Great Plains provinces, New Mexico and western Texas, *in* Mahaney, W. C., ed., Quaternary stratigraphy of North America: Stroudsberg, Pennsylvania; Dowden, Hutchinson, and Ross, Inc., p. 235–274.

Izett, G. A., and Wilcox, R. E., 1982, Map showing localities and inferred distribution of the Huckleberry Ridge, Mesa Falls, and Lava Creek ash beds in the western United States and southern Canada: U.S. Geological Survey Miscellaneous Investigations Map I-1325, scale 1:4,000,000.

Jenny, H., 1941, Factors of soil formation: New York, McGraw-Hill Book Company, 281 p.

Junge, C. E., and Werby, R. T., 1958, The concentration of chloride, sodium, potassium, calcium, and sulfate in rain water over the United States: Journal of Meteorology, v. 15, p. 417–425.

Krynine, D. P., and Judd, W. R., 1957, Principals of engineering geology and geotechnics: New York, McGraw-Hill, 730 p.

Ku, T. L., Bull, W. B., Freeman, S. T., and Knauss, K. G., 1979, Th^{234}-U^{234} dating of pedogenic carbonates in gravelly desert soils of Vidal Valley, southeastern California: Geological Society of America Bulletin, Part 1, v. 90, p. 1063–1073.

Lambert, P. W., 1968, Quaternary stratigraphy of the Albuquerque area, New Mexico: [Ph.D. thesis]: Albuquerque, University of New Mexico, 300 p.

Lamplugh, G. W., 1902, Calcrete: Geological Magazine, v. 9, p. 75.

Lattman, L. H., 1973, Calcium carbonate cementation of alluvial fans in southern Nevada: Geological Society of America Bulletin, v. 84, p. 3013–3028.

Machette, M. N., 1978a, Geologic map of the San Acacia quadrangle, Socorro County, New Mexico: U.S. Geological Survey Geologic Quadrangle Map GQ-1415, scale, 1:24,000.

—— 1978b, Dating Quaternary faults in the southwestern United States using buried calcic paleosols: U.S. Geological Survey Journal of Research, v. 6, p. 369–381.

—— 1978c, Late Cenozoic geology of the San Acacia-Bernardo area, [New Mexico], in Hawley, J. W., compiler, Guidebook to Rio Grande rift in New Mexico and Colorado: New Mexico Bureau of Mines and Mineral Resources Circular 153, p. 135–136.

—— 1982, Guidebook to the late Cenozoic geology of the Beaver basin, south-central Utah: U.S. Geological Survey Open-File Report 82-850, 42 p.

Machette, M. N., Birkeland, P. W., Markos, G., and Guccione, M. J., 1976, Soil development in Quaternary deposits in the Golden-Boulder portion of the Colorado Piedmont, *in* Epis, R. C., and Weimer, R. J., eds., Studies in Geology: Professional Contributions of Colorado School of Mines no. 8, p. 217–259.

Marbut, C. F., 1935, Soils of the United States, Part III, *in* Atlas of American agriculture: U.S. Department of Agriculture, p. 1–98.

McFadden, L. D., and Tinsley, J. C., 1985, Rate and depth of pedogenic carbonate accumulation in soils: Formulation and testing of a compartment model, *in* Weide, D. L. ed., Soils and Quaternary geology of the southwestern United States: Geological Society of America Special Paper 203 (this volume).

Price, W. A., Elias, M. K., and Frye, J. C., 1946, Algal reefs in cap rock of Ogallala Formation on Llano Estacado Plateau, New Mexico and Texas: American Association of Petroleum Geologists Bulletin, v. 30, p. 1742–1746.

Rosholt, J. N., 1980, Uranium-trend dating of Quaternary sediments: U.S. Geological Survey Open-File Report 80-1087, 34 p.

Ruhe, R. V., 1965, Quaternary paleopedology, *in* Wright, H. E., Jr., and Frey, D. G., eds., The Quaternary of the United States, A review volume for the VII Congress of the International Association for Quaternary Research: Princeton, Princeton University Press, p. 755–764.

Scott, G. R., 1963, Quaternary geology and geomorphic history of the Kassler quadrangle, Colorado: U.S. Geological Survey Professional Paper 421-A, 70 p.

—— 1975, Reconnaissance geologic map of the Buena Vista quadrangle, Chaffee and Park Counties, Colorado: U.S. Geological Survey Miscellaneous Field Studies Map MF-657, scale 1:62,500.

Soil Survey Staff, 1975, Soil taxonomy: a basic system of soil classification for making and interpreting soil survey: U.S. Department of Agriculture, Soil

Conservation Service, Agriculture Handbook No. 436, Washington, D.C., U.S. Government Printing Office, 754 p.

Spaulding, W. G., Leopold, E. B., and Van Devender, T. R., 1983, Late Wisconsin paleoecology of the American Southwest, *in* Wright, H. E., Jr. Late-Quaternary environments of the United States, v. 1 (The Late Pleistocene; Porter, S. C., ed.): Minneapolis, University of Minnesota, p. 259–293.

Strain, W. S., 1966, Blancan mammalian fauna and Pleistocene formations, Hudspeth County, Texas: Austin, Texas Memorial Museum Bulletin no. 10, 55 p.

U.S. National Oceanic and Atmospheric Administration, 1978, Climates of the States, with current tables of normals 1941–1970 and means and extremes to 1975: Detroit, Book Tower, v. 1, p. 1–606, v. 2, p. 607–1185.

U.S. Soil Conservation Service, 1970, Soils in the United States, *in* National Atlas of the United States: Washington, U.S. Geological Survey, p. 85–88.

Van Devender, R. R., and Spaulding, W. G., 1979, Development of vegetation and climate in the southwestern United States: Science, v. 204, p. 701–710.

Vanden Hueval, R. C., 1966, The occurrence of sepiolite and attapulgite in the calcareous zone of a soil near Las Cruces, New Mexico: Proceedings, 13th National Conference on Clays and Clay Minerals, v. 25, p. 193–207.

MANUSCRIPT ACCEPTED BY THE SOCIETY JANUARY 12, 1985

Geological Society of America
Special Paper 203
1985

Rate and depth of pedogenic-carbonate accumulation in soils: Formulation and testing of a compartment model

Leslie D. McFadden
Department of Geology
University of New Mexico
Albuquerque, New Mexico 87131

John C. Tinsley
U.S. Geological Survey
345 Middlefield Road
Menlo Park, California 94025

ABSTRACT

The rate and depth of pedogenic carbonate accumulation in soils formed in Quaternary alluvium may be viewed as a theoretical problem that involves the mutual interaction of several independent and dependent soil-forming variables. We propose a model for carbonate accumulation in which the soil column is defined by a vertical sequence of 1-cm^2-area compartments, each with a specified texture, bulk density, water-holding content, lithologic and mineralogic composition, soil-air pCO_2, ionic strength, and temperature. On the basis of these data, rates of carbonate solubility and dissolution within a given compartment are determined. In arkosic to lithic arkosic sandy parent materials, high carbonate solubility (0.137 to 0.212 mg/ml) and the large reactive surface area of eolian calcareous dust result in rapid carbonate dissolution (0.79 to 9.92×10^{-10} g/cm^2/sec) that promotes rapid translocation of carbonate by infiltrating water. We derive a group of equations and use them to calculate net carbonate depletion or accumulation in a soil compartment over an interval of time as a function of the independent variables temperature and precipitation. These two variables largely determine or strongly influence soil-water balance, the external carbonate influx rate, and carbonate solubility.

The carbonate distribution that our model predicts closely resembles the observed carbonate distribution in soils associated with Holocene deposits forming in arid, hyperthermic to xeric, thermic moisture-temperature regimes in southern California. This modeling indicates that with a mean carbonate influx rate of 1×10^{-4} g/cm^2/yr and in a semiarid, thermic climate, the maximum depression of the top of the Cca horizon is attained within only a few thousand years. In contrast, given the same influx rate, our model predicts that a noncalcareous B horizon cannot form in an arid, hyperthermic climate, a conclusion supported by field and laboratory studies of calcic soils in this climate.

The influence of glacial-to-interglacial climatic changes on carbonate accumulation can be modeled by calculating latest Pleistocene soil-water balance with the aid of published estimates of full-glacial temperature and precipitation. On the basis of these modeling results, we propose that either of two types of glacial-to-interglacial climatic changes may account for the strongly bimodal, apparently polygenetic carbonate distribution that is observed in late Pleistocene soils of the eastern Mojave Desert of southern California. Such results of compartment-strategy modeling are encouraging and indicate

the great potential of combined theoretical and empirical methods for considering pedo-
logical problems of interest to Quaternary geologists.

INTRODUCTION

The development of pedogenic horizons containing calcium carbonate is common in arid to semiarid climates. In a classic study, Gile and others (1966) show that pedogenic calcium carbonate (hereinafter referred to as "carbonate") accumulates in increasingly older alluvial deposits and results in a sequence of distinct, morphological stages of calcic horizon development. Quantitative analysis of pedogenic carbonate content can be used to estimate numerical ages of soils (Bachman and Machette, 1977; Machette, 1978; Machette, this volume). The distribution of pedogenic carbonate also can be related to key parameters of climate. For example, the depth at which carbonate accumulates in soils is closely related to mean annual precipitation (Jenny, 1941; Arkley, 1963; Gile, 1975, 1977). These studies demonstrate the significant and diverse applications of pedogenic carbonate accumulation to Quaternary geologic problems.

Birkeland (1974) points out, however, that many factors may affect the distribution of carbonate in soils. The depth of carbonate distribution may be closely related to differences in parent material, texture, and permeability rather than to climatic parameters. Changes in the Ca^{++} influx affect significantly the rate and possibly the depth of carbonate accumulation in soils (Bachman and Machette, 1977; Machette, this volume). Therefore, the distribution of pedogenic carbonate is not related simply to age of parent material or to climatic parameters; instead, it is determined by the complex interaction of several interdependent variables. In this paper, we describe a model that integrates empirical and theoretical methods and we thus predict the pattern of carbonate accumulation in increasingly older soils in different climates. We compare these predictions to late Quaternary calcic soils in order to assess the capability of the model to predict calcic soil development both in constant and in changing climates.

THE COMPARTMENT MODEL

A compartment model strategy is used because it is versatile, flexible, and accommodates continuously changing values among the interdependent variables that influence soil development. Numerous compartment models have been used to assess soil development (Arkley, 1963; Kline, 1973; Frissel and Reineger, 1974; Rogers, 1980). In a compartment model, the soil column is represented by a vertical sequence of compartments of equal thickness. Each compartment has a cross-sectional area of 1 cm^2 and is characterized by quantitatively determined values of texture, bulk density (D), and mineralogic composition. The first two parameters are used to calculate compartment porosity and available water-holding capacity (AWC). In addition, the percentage of CO_2 in the soil air (pCO$_2$) and temperature are specified for each compartment. The designated soil temperature is

estimated from the mean monthly air temperature and, for simplicity, is considered to be constant with increasing depth.

The procedure used to calculate carbonate gains or losses in a sequence of compartments is given in the following steps:

1. Calculate the total annual volume of water (Vw) that leaches the uppermost calcareous compartment:

$$Vw = Li - AWC_t \qquad (1)$$

Li (leaching index) = effective soil annual moisture per 1 cm^2 column (Arkley, 1963), and AWC$_t$ = the sum of the field-capacity water content minus the permanent-wilting-point water contents of the uppermost calcareous compartment and of all superjacent noncalcareous compartments.

2. Calculate the total mass of carbonate (Md) annually dissolved by Vw at a temperature and a pCO$_2$ designated for each compartment:

$$Md = Ms\ Vw \qquad (2)$$

where Ms = carbonate solubility, in g/cm^3.

3. Calculate the annual rate of depletion (Rd) of carbonate in the uppermost calcareous compartment:

$$Rd = Me + Md \qquad (3)$$

where Me = total mass of eolian carbonate annually accumulated in the soil. Note: The model as formulated in this paper does not consider the condition in which Me is greater than Md. This condition would result in the uppermost compartment not becoming depleted in carbonate.

4. Calculate the time (Td) required to deplete parent-material carbonate (cp) from the uppermost calcareous compartment:

$$Td = \frac{cp}{Rd} \qquad (4)$$

5. Calculate the annual rate of accumulation of secondary carbonate (cs/yr) in all lower compartments (1, 2 . . . n):

$$cs/yr_{(1,\,2\ldots n)} = Ms_{(1,\,2\ldots n)}\ AWC_{(1,\,2\ldots n)} \qquad (5)$$

6. For each compartment 1, 2 . . . n, calculate the mass of carbonate accumulation (cs$_{t(1,\,2\ldots n)}$) for the interval of time Td:

$$cs_{t(1,\,2\ldots n)} = (Td)(cs/yr)_{(1,\,2\ldots n)} \qquad (6)$$

Equations (1) through (6) are valid if compartments in the zone of carbonate accumulation are characterized by values of

pCO_2 that are constant with depth. If pCO_2 varies with depth, solubility for a given compartment varies also, requiring modification of equation (5). If pCO_2 decreases in successively lower compartments characterized by carbonate accumulation, secondary carbonate accumulation in a given compartment (cs, compartment n) relative to superjacent compartment (n–1) is given as:

$$cs_1 = (Ms_2 - Ms_1)(Vw - AWC^*) + (Ms_2)(AWC_1) \qquad (7)$$

where AWC^* = the field capacity minus permanent-wilting-point water content (AWC) of the nth compartment plus the sum of the AWC of all superjacent compartments that are present below the uppermost calcareous compartment, except for the lowest compartment where $cs_1 = (Ms_2)(AWC_1)$.

If pCO_2 increases in successively lower compartments, carbonate depletion will occur in those compartments as discussed later in this paper. For this case, the rate of carbonate depletion for the nth compartment (Rd_n) relative to superjacent compartment (n–1) is given as:

$$Rd_n = - (Ms_n - Ms_{(n-1)})(Vw - AWC^*) \qquad (8)$$

Total gains or losses for given compartments over time Td may now be calculated by substituting values of cs_1 or Rd_1 as derived in equations (7) or (8) for cs_t in equation (6). For the entire soil column, at time Td, the absolute increase in pedogenic carbonate would be equal to (Me)(Td). This value is defined as the whole-profile index of secondary carbonate (cS) (see Machette, this volume).

Evaluation of Compartment Model Variables

The values of compartment AWC, D, and porosity depend on compartment texture. Each of these physical characteristics changes significantly as soils progressively develop; these changes require periodic recalculation of compartment parameters. Studies of soil chronosequences in different climates (see Harden and Marchand, 1977; Gile and Grossman, 1979; and McFadden, 1982) provide the data necessary to estimate realistic time-dependent changes in compartment parameters subsequent to the time of initial soil development (t_0). This procedure permits us to predict the influence of variables such as climate on carbonate accumulation over several tens of thousands of years.

The procedure outlined above in equations (1) through (6) entails the evaluation of carbonate solubility (Ks) as well as the estimation of carbonate dissolution rates (R) in soil environments. The values of these variables are determined by independent variables of temperature and precipitation which also determine annual soil moisture balance. Another important independent variable in the model, eolian carbonate influx (Me) must be evaluated (see Machette, this volume). In the following sections, we discuss the methods used to estimate soil moisture balance,

carbonate solubility, and rates of carbonate dissolution and Ca^{++} influx in open-system soil environments.

Soil-Water Balance

Soil-water balance is determined from climatic and available water-holding capacity (AWC) data. Available water-holding capacity can be calculated empirically from textural and bulk-density data as described by Birkeland (1974) or it can be measured in the field using a lysimeter apparatus (Jenny, 1980). The leaching index (Li) defined by Arkley (1963) approximates the amount of soil moisture available for soil reactions. The mean monthly Li values in a given area are calculated from monthly temperature, precipitation, and potential evapotranspiration (ETp) data. Temperature and precipitation data are readily available for many areas. ETp is estimated using the Thornthwaite method as described by Palmer and Havens (1958).

The mean monthly infiltration depth is determined jointly from Li and AWC data, where infiltration is assumed to occur as a uniform wetting front that recharges successively lower soil compartments (1, 2 . . . n) to field capacity. (As defined in Birkeland (1974), field capacity is the water content at which forces holding water films on particle surfaces equal the force of gravity.) Gravel has no available water-holding capacity; thus, the absolute infiltration depth (Depth*) for a soil composed of 50 percent gravel corresponds to twice the depth determined for the nongravelly fraction. Furthermore, capillary rise is assumed to be zero for highly permeable, gravelly soils (Birkeland, 1974, Machette, this volume).

Figure 1 shows the calculated soil-water balance for texturally homogeneous gravelly sandy soils in three climatic regimes of southern California and southwestern Arizona (see Table 1): arid, hyperthermic; semiarid, thermic; and xeric, thermic. Data in Table 1 and Figure 1 indicate that leaching of soil profiles in all three climates occurs only during the winter season owing to the combination of high evapotranspiration and low precipitation that occurs during the remaining part of the year.

Carbonate Solubility

The solubility of carbonate in the system $CaCO_3$ - CO_2 - H_2O at a given pCO_2 is calculated by simultaneous solution of a suite of equations that describe carbonate equilibria (Garrels and Christ, 1965). Drever (1982) gives the following equation that expresses the molality (m^3) of Ca^{++} as a function of pCO_2:

$$m^3Ca^{++} = \frac{pCO_2\, K_1\, K_c\, K_{CO_2}}{4\, K_2\, \gamma\, Ca^{++}\, \gamma^2\, HCO_3^-} \qquad (9)$$

where K_c, K_1, K_2 and K_{CO_2} are dissociation constants in the carbonate system and γCa^{++} and γHCO_3^- are the activity coefficients of Ca^{++} and HCO_3^-.

Values of the dissociation constants can be calculated over the temperature range 0° to 50°C, thereby permitting calculation of carbonate solubility as a function of temperature for a given

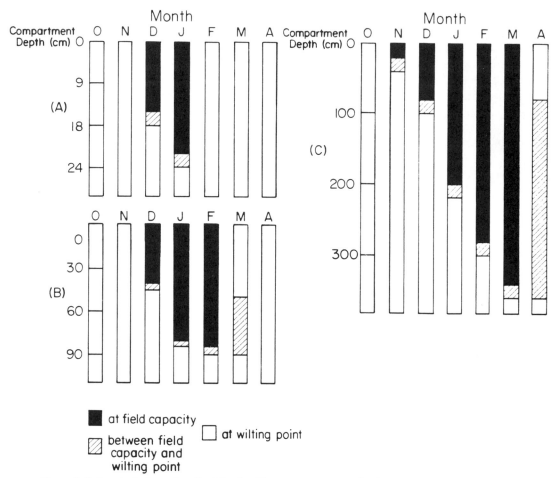

Figure 1. Soil-water balance for alluvial soils of homogenous texture for October through April. During the remainder of the year, soils are at permanent wilting point throughout the column. (A) Arid, hyperthermic regime, AWC = 0.12 ml/cm^3; (B) Semiarid, thermic regime, AWC = 0.04 ml/cm^3; (C) Xeric, thermic regime, AWC = 0.05 ml/cm^3.

pCO$_2$ (Jacobson and Langmuir, 1974). Picknett and others (1976) and Bogli (1980) present carbonate solubility data for the range of temperatures and pCO$_2$ conditions that are present in soils.

For soils in which carbonate is the only readily soluble mineral, soil pCO$_2$ is low (less than $10^{-3.0}$), and soil-water ionic strength (I) is less than $10^{-3.0}$, the value of both activity coefficients is approximately 1. Increasingly higher soil pCO$_2$ (greater than $10^{-3.0}$) and the presence of other, relatively soluble minerals such as gypsum or the presence of dissolved Mg^{++} considerably influence carbonate solubility (Krauskopf, 1967; Plummer and Wigley, 1976). In general, these solution conditions act to increase soil-water ionic strength. According to the Debye-Huckel theory, increases in ionic strength result in decreases in the values of γCa^{++} and γHCO$_3^-$, which causes an increase in the solubility of carbonate, as shown in equation (9) (Krauskopf, 1967; Drever, 1982).

Ionic strength can be determined for soil water from electroconductivity data derived from field or laboratory measurements (Griffin and Jurinak, 1973). These data, combined with soil mineralogic, temperature, and pCO$_2$ data, permit calculation of carbonate solubility in soils forming under a wide range of climatic conditions.

Several studies show that many aqueous environments, including that of soil water, apparently are supersaturated with respect to carbonate (Barnes, 1965; Suarez, 1981; Suarez and Rhoades, 1982). For example, Suarez and Rhoades propose that anorthite, rather than calcite, may control Ca^{++} activity in fine, loamy calcareous soils after two weeks to two months of reaction time, thereby resulting in apparent carbonate supersaturation. However, as discussed later, rapid soil infiltration rates and the rapid dissolution rates of calcite relative to aluminosilicates of sandy, arkosic soils may well favor control of soil carbonate on Ca^{++} activity during periods of leaching.

Carbonate precipitation probably is favored by two factors in soils forming in arid to semiarid climates: (1) fine particles suspended in soil water are ubiquitous in natural systems; the presence of fine particulates favors carbonate precipitation as the amounts of surface free energy required for nucleation are decreased, and (2) high evapotranspiration during the summer sea-

TABLE 1. TEMPERATURE, PRECIPITATION, POTENTIAL EVAPOTRANSPIRATION (ET$_p$), AND LEACHING INDEX (Li) DATA FOR SELECTED SITES IN SOUTHERN CALIFORNIA AND SOUTHEASTERN ARIZONA CHARACTERIZED BY ARID TO XERIC SOIL-MOISTURE AND HYPERTHERMIC TO THERMIC SOIL-TEMPERATURE REGIMES[1].

	J	F	M	A	M	J	J	A	S	O	N	D
Parker Reservoir, California				Altitude: 226 m		Li: 0.97 cm						
Temperature[2]	11.60	14.30	17.30	22.00	26.80	31.30	35.10	34.40	31.30	24.90	17.10	12.40
Precipitation[3]	1.40	1.02	1.45	0.71	0.13	0.05	0.76	1.35	0.99	0.76	1.07	1.40
ET$_p$[4]	0.71	2.36	3.81	-	-	-	-	-	-	-	3.73	1.12
Palm Springs, California				Altitude: 130 m		Li: 2.25 cm						
Temperature	12.40	14.70	16.80	20.70	24.40	28.30	32.70	31.90	29.00	23.50	16.90	12.80
Precipitation	2.87	1.68	1.60	0.61	0.10	0.00	0.61	0.61	1.42	0.51	1.73	2.57
ET$_p$	1.50	2.23	-	-	-	-	-	-	-	-	3.59	1.69
Barstow, California				Altitude: 660 m		Li: 1.36 cm						
Temperature	7.60	10.20	12.60	16.70	21.00	25.30	29.40	28.30	24.80	18.70	12.00	7.80
Precipitation	1.60	1.24	1.32	0.61	0.10	0.10	0.81	0.76	0.61	0.66	1.22	1.75
ET$_p$	0.98	1.63	-	-	-	-	-	-	-	-	-	1.01
Beaumont, California				Altitude: 800 m		Li: 18.6 cm						
Temperature	8.80	9.60	10.50	13.30	16.30	19.90	24.60	24.80	22.30	18.70	12.90	9.80
Precipitation	7.21	7.01	6.58	4.19	1.04	0.23	0.51	0.58	0.76	1.52	5.00	6.32
ET$_p$	1.73	2.13	3.05	4.23	-	-	-	-	-	-	3.51	3.15
Arroyo Seco Canyon, California				Altitude: 400 m		Li: 30.3 cm						
Temperature	11.40	12.30	13.20	14.90	16.80	18.80	22.30	22.70	21.80	18.60	14.90	12.20
Precipitation	11.80	12.70	8.66	4.78	1.09	0.33	0.08	0.23	0.81	2.98	4.24	10.10
ET$_p$	2.59	3.15	4.14	-	-	-	-	-	-	-	4.35	3.05

[1]ET$_p$ values are shown only for months when ET$_p$ is less than or slightly exceeds precipitation. Temperature, precipitation data from National Oceanic and Atmospheric Administration (1978).
[2](OC), mean monthly.
[3](cm), mean monthly.
[4](cm), monthly.

son greatly depletes soil moisture and results in supersaturation of soil moisture and subsequent deposition of carbonate.

Equation (9) shows that increasing pCO$_2$ increases carbonate solubility. A mean atmospheric pCO$_2$ of $10^{-3.5}$ atm determines the initial content of dissolved CO$_2$ in rainwater. Plant respiration and oxidation of organic matter increase soil pCO$_2$ by a factor of five to as much as two orders of magnitude times the atmospheric value in poorly drained, aquatic soils (Baver, 1956; Jenne, 1971; Garrels and MacKenzie, 1971; Bohn and others, 1979; Lindsay, 1979). Tinsley and others (1981) measured maximum pCO$_2$ levels of about $10^{-3.4}$ atm in coniferous forest soils during the late summer in the southern Sierra Nevada, California. Parada and others (1983) showed that pCO$_2$ value in soils forming in a semiarid climate near Tucson, Arizona, also varies seasonally; pCO$_2$ values of $10^{-2.0}$ to $10^{-2.5}$ atm are characteristic of the winter season. There is, however, a paucity of soil pCO$_2$ data for soils forming in different environments. We have estimated pCO$_2$ values for these soils based on published known maximum and minimum soil pCO$_2$ values, the climate, the vegetation that characterizes the climate, and soil data including organic-matter content, pH, and root and pore density. Table 2 shows soil pCO$_2$ and corresponding values of calcite solubility estimated for three climates at different temperatures.

Soil pCO$_2$ presumably varies with depth as a result of changes in root density, organic matter content, pore geometry, and other factors. In well-aerated soil profiles in which soil moisture is at or near the permanent-wilting stage, rapid rates of upward diffusion of CO$_2$ in air may cause decreases in soil pCO$_2$ with increasing depth. This trend may become more pronounced below the zone of maximum biotic activity. In other soils, the soil pCO$_2$ values may increase downward (Blatt and others, 1981). The highest soil pCO$_2$ values may occur near the base of or possibly below the A horizon of the soil, owing to CO$_2$ respiration associated with deep root systems and to relatively low rates of CO$_2$ diffusion in water in soils with moisture at or near field capacity. For example, Petraitus and Wood (1982) reported that in the Ogallala aquifer system pCO$_2$ progressively increased downward to a maximum of one hundred times the atmospheric value. The pattern of carbonate accumulation with depth as a function of changing values of soil pCO$_2$ may be considered by using equations (8) and (9) and comparing these results to those obtained for soils in which pCO$_2$ is assumed to be relatively constant with depth.

Carbonate Influx Rates

Possible sources of pedogenic carbonate include parent-material carbonate, weathering of Ca^{++}-bearing aluminosilicate minerals, and external sources such as windblown calcareous dust or Ca^{++} dissolved in precipitation (see Machette, this volume). Calcic soils modeled in this study are formed in parent materials that are noncalcareous or weakly calcareous (2 to 5 percent) arkosic sediments, so primary carbonate constitutes only a minor source of pedogenic carbonate. Hydrolytic weathering of parent materials also can provide minimal amounts of pedogenic carbonate. For example, to derive the mass of pedogenic carbonate that has accumulated in arid, Holocene soils of southern California or in Pleistocene soils of southern Nevada by means of hydrolytic weatheing would require extensive chemical weathering of parent material; in fact, very little evidence of chemical weathering of these parent materials is observed (Gardner, 1972; McFadden, 1982). Therefore, the principal source of pedogenic carbonate accumulated in noncalcareous to weakly calcareous parent materials must be external (Gile and others, 1966; Machette, this volume). Pedogenic carbonate accumulation rates vary significantly in the western United States, probably in response to climate and airborne Ca^{++} influx (Bachman and Machette, 1977; Machette, this volume). For simplicity, in this study a constant influx is assumed owing to lack of data in southern California concerning airborne Ca^{++} influx with time. The changing rate of carbonate influx may be modeled to determine theoretically the impact of this factor on the rate and depth of carbonate accumulation.

Theoretical calculations show that average rainwater contains less than 0.1 ppm Ca^{++} (Garrels and MacKenzie, 1971). Data from Junge and Werby (1958) suggest that precipitation over the Mojave Desert, southern California, possesses from 0.5 to 2.0 ppm Ca^{++}. Thus, rainwater is typically highly undersaturated with respect to carbonate and any Ca^{++} in rainwater has a negligible effect on carbonate solubility. Over very long periods of time, Ca^{++} in rainwater could account for a substantial amount of pedogenic carbonate (Gardner, 1972; Bachman and Machette, 1977), particularly if rainwater maintains a Ca^{++} content of several ppm over long periods of time. Temporal variation in the Ca^{++} in rainwater during the past is unknown. For example, given rainwater containing a Ca^{++} concentration of 1.5 ppm, an arid climate with 10 cm of annual precipitation, and a total Ca^{++} influx of 1×10^{-4} g/cm^2/yr, Ca^{++} in rainwater could account for 15 percent of the pedogenic carbonate accumulated annually. This suggests that for most of the Mojave Desert, the majority of Holocene pedogenic carbonate is derived from carbonate dust, assuming an average rainwater Ca^{++} content of 1.5 ppm during many thousand years. In much wetter climates, of course, Ca^{++} in rainwater could account for significantly more of the pedogenic carbonate. In hot, arid to semiarid climates, however, most of the Ca^{++} in rainwater falling duirng spring through autumn may precipitate annually as small amounts of carbonate in the uppermost part of the soil. This postulated annual accumulation of carbonate plus the annually entrapped carbonate dust very closely approximates Me as defined in equation (3).

Carbonate Dissolution Rates

In addition to the controls of carbonate solubility, the maximum mass of carbonate that can be transported during an infiltration event is determined by the rate of carbonate dissolution. The morphology of pedogenic carbonate suggests that carbonate dissolution rates are relatively rapid compared to those of most aluminosilicates. Pedogenic carbonate is typically present as 1-μm particles (micrite) indicating the presence of crystallites and

crystallization from an aqueous phase (Sehgal and Stoops, 1972; Bachman and Machette, 1977). Such evidence indicates that carbonate translocation occurs in solution rather than by mechanical transport or suspension (Soil Survey Staff, 1975). However, the solubility of some Ca-bearing aluminosilicates may approach or even exceed the solubility of carbonate in soil environments (Suarez and Rhoades, 1982). Thus, a quantitative evaluation of the range of rates of carbonate dissolution relative to aluminosilicate dissolution in soil environments is needed to evaluate the magnitude of carbonate translocation relative to hydrolysis-derived Ca^{++} during soil leaching events.

Most studies of calcite dissolution rates show that calcite dissolution is a rapid, surface-controlled reaction (Reddy and Nancollas, 1971; Plummer and Wigley, 1976; Plummer and others, 1978). Dreybrodt (1981), based on the general rate equation determined by Plummer and others, derived rate equations for systems that are either open or closed to atmospheric pCO_2 as a function of carbonate surface area (F) and fluid volume (V). For the open system, Dreybrodt showed that the conversion of CO_2 to H_2CO_3 is the rate-limiting step when the ratio of the volume of solution to calcite surface area (V/F) is less than 0.01 cm. Apparently, the rapid surface-controlled rate of dissolution quickly depletes available H_2CO_3; thus the open system dissolution rate is given as:

$$R = \frac{V}{F} (kCO_2)(CO_2) \tag{10}$$

where kCO_2 and CO_2 may be calculated for a given temperature and specified pCO_2. The dissolution rate of carbonate calculated using equation (10) is the probable minimum dissolution rate in open systems that we consider to characterize well-aerated, permeable soils. Table 3 shows carbonate dissolution rates calculated for different temperatures, and pCO_2 at a V/F ratio of 0.001 cm. This value is obtained for gravelly, loamy-sandy soils in which carbonate is present chiefly as particles having diameters less than 0.625 mm.

Data given in Tables 2 and 3 permit us to estimate the time required to saturate a soil-wetting front with respect to soil carbonate at a specified temperature and soil pCO_2. For example, consider a parent material that has the following properties: a loamy-sandy texture, carbonate content of 5.0 percent, D = 1.70 g/cm^3, AWC = 0.15 ml/cm^3, average porosity of 36.0 percent, mean particle size = 1×10^{-4} cm, and reactive surface area = 7.35×10^2 cm^2/cm^3. Assuming an infiltration rate of 60 cm/hour (very rapid permeability), carbonate equilibration should theoretically occur at a depth as shallow as 3.2 cm in a period of 192 seconds, where soil temperature = 12°C and soil $pCO_2 = 10^{-2.5}$ atm. With more rapid infiltration rates and finer carbonate particle size, carbonate equilibration would occur at even shallower depths and in a much shorter period of time.

In contrast to the rapidity of carbonate dissolution, dissolution rates of most aluminosilicates (Busenberg and Clemency, 1976; Siever and Woodford, 1979) which are abundant in arkosic sediments, are several orders of magnitude slower than the

TABLE 2. EXAMPLES OF CALCITE SOLUBILITY AS FUNCTIONS OF pCO_2, TEMPERATURE, AND IONIC STRENGTH (I)
ESTIMATED FOR DIFFERENT SOIL-MOISTURE REGIMES.*

Temp. (°C)	pCO_2 (atm)	CO_2^0 (ppm)	pH	$CaCO_3$ (ppm)	$CaCO_3$ (mg/ml)	I (mol/l)	γM^{++}	$(\gamma M^{++2})^{-1}$	Solubility[1] (mg/ml)
				ARID					
12	$10^{-3.3}$	1.1	8.3	68	0.068	7.8×10^{-4}	0.88	1.29	0.088
12	$10^{-2.8}$	3.3	7.7	106	0.106	7.8×10^{-4}	0.88	1.29	0.137
9	$10^{-2.5}$	7.6	7.4	130	0.130	7.8×10^{-4}	0.88	1.29	0.168
4	$10^{-2.5}$	9.3	7.5	147	0.147	7.8×10^{-4}	0.88	1.29	0.190
				SEMIARID					
12	$10^{-2.5}$	6.9	7.6	125	0.125	4×10^{-4}	0.91	1.21	0.151
9	$10^{-2.0}$	23.0	7.2	183	0.183	4×10^{-4}	0.91	1.21	0.221
4	$10^{-2.0}$	29.0	7.2	212	0.212	4×10^{-4}	0.91	1.21	0.257
				XERIC					
12	$10^{-2.0}$	20.0	7.4	175	0.175	4×10^{-4}	0.91	1.21	0.212

*CO_2^0, pH, and calcite solubility data from Picknett and others (1976).
[1]Solubility corrected for I.

TABLE 3. EXAMPLES OF CALCITE DISSOLUTION RATES AS FUNCTIONS OF
TEMPERATURE AND pCO_2 IN DIFFERENT MOISTURE REGIMES

Temp. (°C)	pCO_2 (atm)	CO_2^0 (mol/ml)	k_{CO} (sec^{-1})	$CaCO_3$ Dissolution Rate[1] (mol/cm^2/sec)	(g/cm^2/sec[1])
			ARID		
12	$10^{-3.3}$	2.50×10^{-8}	1.0×10^{-2}	2.5×10^{-13}	2.5×10^{-11}
12	$10^{-2.8}$	7.86×10^{-8}	1.0×10^{-2}	7.9×10^{-13}	7.9×10^{-11}
9	$10^{-2.5}$	1.76×10^{-7}	5.3×10^{-3}	9.3×10^{-11}	9.3×10^{-11}
4	$10^{-2.5}$	1.95×10^{-7}	1.5×10^{-3}	7.6×10^{-13}	7.6×10^{-11}
			SEMIARID		
12	$10^{-2.5}$	1.56×10^{-7}	1.0×10^{-2}	1.6×10^{-12}	1.6×10^{-10}
9	$10^{-2.5}$	5.56×10^{-7}	5.3×10^{-3}	3.0×10^{-12}	3.0×10^{-10}
4	$10^{-2.5}$	6.17×10^{-7}	1.5×10^{-3}	9.3×10^{-13}	9.3×10^{-10}
			XERIC		
12	$10^{-2.0}$	4.96×10^{-7}	1.0×10^{-2}	9.92×10^{-12}	9.92×10^{-10}

[1]$R_{Calcite} = {}^{k}CO_2(pCO_2) \frac{V}{F}$; $\frac{V}{F} = 0.001$; k_{CO_2} is rate constant in $CaCO_3$-CO_2-H_2O
system.

minimum rates determined for carbonate, especially at soil pH values exceeding 7.0. For example, Busenberg and Clemency calculate a linear (zero order) dissolution rate of 2.40×10^{-16} mol/cm^2/sec for anorthite, a value that is typical of rates determined for other feldspars. Reactive surface area significantly influences dissolution rates; in sandy, gravelly parent materials, particle sizes of most aluminosilicates (coarse silt to coarse sand) are much larger than the estimated mean particle diameter of eolian calcareous dust, a difference in surface reactive areas of one to two orders of magnitude. Thus, during an infiltration event, it is unlikely that aluminosilicate-soil water reactions significantly influence carbonate dissolution rates and subsequent carbonate mass transfer, given slow dissolution rates of abundant aluminosilicates, neutral or higher soil pH, and relative differences in surface reactive area. It also seems probable, given equal net infiltration, that a large infiltration event or several small infiltration events will dissolve and transport similar masses of carbonate as carbonate equilibrium is achieved rapidly following an initial condition of strong undersaturation.

PREDICTED PATTERNS OF PEDOGENIC CARBONATE ACCUMULATION IN HOLOCENE DEPOSITS

Although climatic fluctuations have occurred during the Holocene, these fluctuations apparently have been small compared with the difference between interglacial and glacial climates. Thus, to model the development of Holocene calcic soils, we have assumed a "constant" climate. Although the Holocene-Pleistocene boundary is placed at 10,000 years B.P. by many

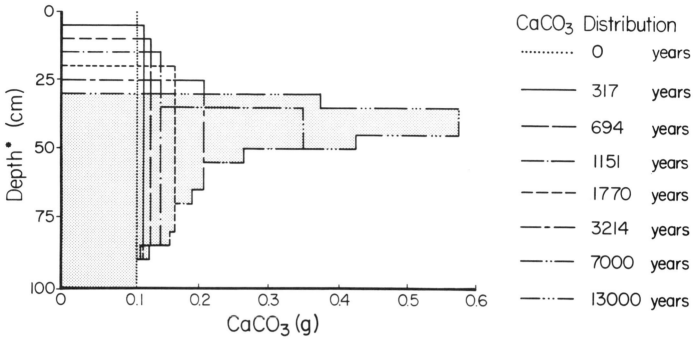

Figure 2. Predicted pattern of Holocene pedogenic carbonate accumulation in a semiarid, thermic climate (Li = 3.5 cm; external carbonate influx rate = 1.5×10^{-4} g/cm^2/yr; pCO$_2$ = $10^{-2.5}$ atm). Dotted line shows initial carbonate distribution at t$_o$. Different line patterns show calculated carbonate distribution after indicated duration. Gray area indicates distribution of carbonate after 13,000 years. Depth* = absolute infiltration depth in the <2 mm size fraction.

geologists (see, for example, Hopkins, 1975; Sohl and Wright, 1980) botanical and paleontologic evidence from the southwestern United States indicate that the initial change towards a warm "Holocene" climate occurred well before 10,000 years B.P. and as early as 13,000 to 14,000 years B.P. (Van Devender, 1973, 1977; Spaulding, 1982). Using these data, we predict patterns of calcic soil development for the Holocene and we use a 13,000-year period for the purpose of our report.

Semiarid, Thermic Climate

The patterns of Holocene carbonate accumulation in gravelly, sandy alluvium predicted by the compartment model in a semiarid, thermic climate are shown in Figures 2 and 3. For comparative purposes, Figure 4 shows profile data for soils formed in latest Holocene and in middle Holocene deposits in southern California that have formed in this climate. The observed distribution of carbonate exhibited by these soils is similar to the predicted distribution of carbonate in both the middle and the late Holocene calcic soils. Profile descriptions of soils present in latest Pleistocene(?) to early Holocene(?) deposits forming in this climate in southern California indicate that the depth to the top of the zone of carbonate accumulation is about 60 cm (McFadden, 1982). This depth is close to the depth of 40 to 50 cm predicted by the model for a Holocene maximum depth of accumulation in gravelly soils shown in Figures 2 and 3.

The modeling also suggests several important aspects of the

rates and processes of carbonate accumulation in a semiarid, thermic climate: (1) Carbonate depletion at shallow depths is initially very rapid. The rate of depletion decreases exponentially, however, with increasing soil age. The maximum depth to the top of the zone of carbonate accumulation is attained theoretically in only a few thousand years. (2) In early Holocene soils, carbonate may accumulate in the B horizon as a result of decreases in infiltration depth which are due to the accumulation of pedogenic clay in weak argillic B horizons. Deposition of carbonate in the lower part of the B horizon is not uncommon in many Holocene and late Pleistocene calcic soils (Birkeland, 1974; Soil Survey Staff, 1975; Gile and Grossman, 1979; McFadden, 1981, 1982; Machette, this volume). (3) Changing soil pCO$_2$ slightly alters the pattern of carbonate distribution especially during the initial pedogenesis; thus changes in pCO$_2$ result in a slightly different predicted pattern of carbonate distribution (compare Figures 2, 3A, and 3B). As might be expected, decreases in pCO$_2$ below 25-cm depth increase the mass of carbonate deposited in the upper part of the soil; increases in pCO$_2$ from the surface to a maximum at depth of 25 cm significantly increase the rate of carbonate depletion from the upper 25 cm and slightly decrease the maximum Holocene depth to the top of the zone of carbonate accumulation.

Gile and his colleagues have studied development of calcic soils formed in Holocene noncalcareous alluvial fan and terrace deposits in the arid to semiarid, thermic climate of the Basin and Range area of southern New Mexico (Gile, 1975, 1977; Gile and

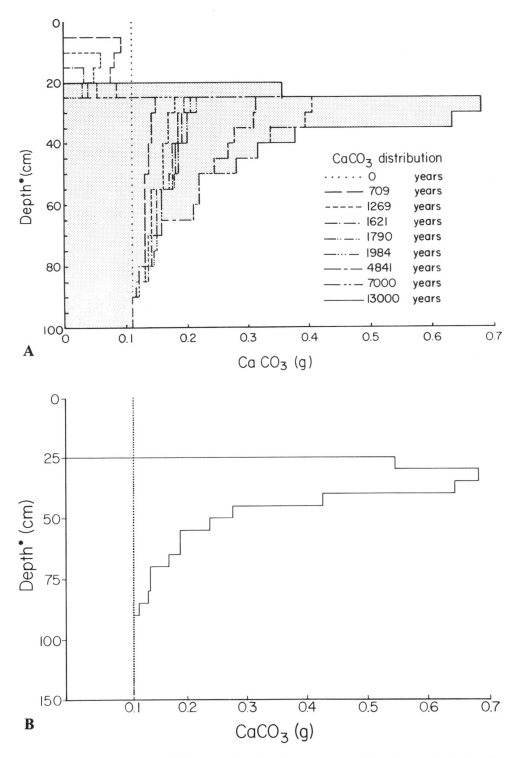

Figure 3. Predicted patterns of Holocene pedogenic carbonate accumulation in a semiarid, thermic climate (Li = 3.5 cm; external carbonate influx rate = 1.5×10^{-4} g/cm^2/yr). (A) pCO$_2$ = $10^{-3.3}$ atm in compartment 1 (0 to 5 cm), increasing to pCO$_2$ = $10^{-2.5}$ atm in compartment 5 (20 to 25 cm). Below compartment 5 (25 cm), pCO$_2$ decreases to minimum value of $10^{-4.0}$ atm in compartment 20 (95 to 100 cm). Dotted line shows initial carbonate distribution at t$_0$. Gray area indicates final distribution of carbonate. (B) Predicted carbonate distribution after duration of 13,000 years, where pCO$_2$ = $10^{-2.5}$ atm at 0 to 25 cm, decreasing below compartment 5 (25 cm) to minimum pCO$_2$ = $10^{-3.5}$ atm in compartment 16 (75 to 80 cm). Depth* = absolute infiltration depth in the <2 mm size fraction.

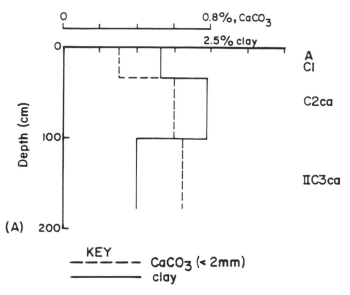

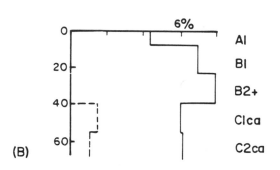

Figure 4. Carbonate and clay content (percent) for soils formed on Holocene deposits in a semiarid, thermic climate in southern California (data from McFadden, 1982). Parent material is gravelly to bouldery lithic arkosic, very weakly calcareous sand to loamy sand. (A) Soil formed on latest Holocene (<1,000 years B.P. (?)) terrace deposits of Mission Creek, eastern San Bernardino Mountains. Pedogenic carbonate also occurs on gravel as thin (<0.1 mm) discontinuous Stage I coatings. (B) Soil formed on middle Holocene (4,000 to 7,000 (?) years B.P.) terrace deposits of the Rodman Mountains, central Mojave Desert. Pedogenic carbonate also present as thin to moderately thick (<0.5 mm) Stage I coatings on gravel.

Grossman, 1979). The depth to the top of the zone of carbonate accumulation in these soils varies considerably; however, in the soil present in the gravelly to very gravelly sediments comprising the Organ alluvium estimated to be 2 to 6 thousand years old, the depth to the Cca horizon is 51 cm. Again, the depth observed in the field is similar to the depth predicted by the model.

Arid, Hyperthermic Climate

In marked contrast to predicted profile development in a semiarid climate is the pattern of carbonate accumulation pre-

dicted for soils forming in an arid, hyperthermic climate (see Fig. 5). A noncalcareous horizon does not form above a zone of carbonate accumulation. Instead, carbonate accumulation is confined to thin near-surface soil horizons. Figure 6 shows laboratory data for soil profiles formed in very gravelly middle Holocene deposits in an arid, hyperthermic climate in the eastern Mojave Desert, southern California. The observed carbonate distribution is similar to that predicted by the model for a middle Holocene profile (see Fig. 5). Similar Holocene soils from the same climate in southwestern Arizona were formed in noncalcareous to weakly calcareous parent materials (Schenker, 1978; McHargue, 1981).

The trend in carbonate accumulation clearly results from the insufficient amount of soil moisture required to transport externally derived carbonate to a subsurface horizon. The condition is defined as "rainfall-limited" by Machette (this volume). The net accumulation of carbonate in soil surface horizons significantly affects other pedogenic processes. For example, the presence of abundant carbonate causes colloid flocculation and significantly reduces the amount of translocation of fine eolian clay. B horizon development in this climate is confined to thin (less than 5 cm thick) zones in the moist microenvironment beneath large surface clasts (McFadden and Bull, 1981). Pedogenic carbonate in the Cca horizon is present chiefly as thin (0.1 to 0.5 mm), discontinuous Stage I coatings (McFadden, 1982). In contrast, pebble coatings are often thicker, more continuous, and occur over a greater horizon thickness in Holocene soils present in southern New Mexico. This may be due to the greater carbonate influx or higher rainfall of southern New Mexico compared to southeastern California (see Machette, this volume). Figure 5B shows the predicted patterns of carbonate accumulation for a carbonate influx 50 percent greater than the influx used to predict the patterns shown in Figure 5A. The increased influx significantly accelerates carbonate accumulation in the upper Cca horizon and Avca horizon. A similar increase in carbonate influx in a moister, semiarid climate would potentially accelerate subsurface calcic horizon development. An increase in influx by a factor of 3 or 4 might decrease the depth to the top of the zone of carbonate accumulation because the maximum carbonate solubility and available soil moisture would be insufficient to transport the significantly increased mass of carbonate to the depth established for the much lower influx rate. This condition again would qualify as moisture-limited (see Machette, this volume).

Xeric, Thermic Climate

Figure 7 shows the pattern of carbonate accumulation predicted for Holocene soils forming in the xeric, thermic climate characteristic of the piedmonts of the Transverse Ranges in southern California. The essentially noncalcic soil profile predicted by our model is consistent with the many field and laboratory studies of soil development in a Mediterranean-type climate in California (Arkley, 1962a, 1962b, 1964; Janda and Croft, 1967; Harden and Marchand, 1977; McFadden and others, 1980;

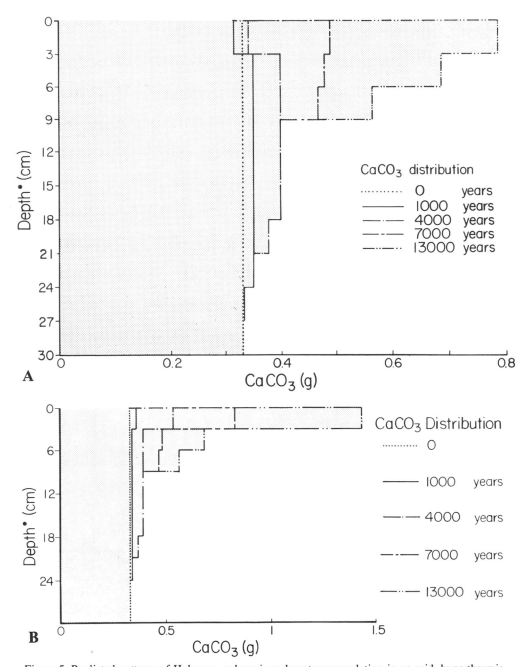

Figure 5. Predicted patterns of Holocene pedogenic carbonate accumulation in an arid, hyperthermic climate (Li = 0.97 cm; $pCO_2 = 10^{-2.8}$ atm). Dotted line shows initial carbonate distribution at t_o. Gray area shows final carbonate distribution. (A) Carbonate influx rate = 1.0×10^{-4} g/cm^2/yr. (B) Carbonate influx rate = 1.5×10^{-4} g/cm^2/yr. Depth* = absolute infiltration depth in the <2 mm size fraction.

Marchand and Allwardt, 1981; McFadden, 1982; Machette, this volume). For example, Figure 8 shows laboratory data for soils that have formed in upper Holocene to lower Holocene arkosic, gravelly fan deposits of the central San Bernardino Mountains of southern California. Carbonate has not accumulated in these gravelly deposits, while clay and iron oxyhydroxides have continued to increase with time, because large amounts of winter moisture leach the profile. Leaching reflects the increased solubil-

ity of carbonate in cool, winter precipitation at the estimated high values of soil pCO_2 (see Tables 1 and 2). These high soil pCO_2 values reflect the observed thick, Mollic epipedons of Holocene and late Pleistocene soils that support many fine to medium roots and contain as much as 1.5 to 5.0 percent organic carbon (McFadden, 1982).

Figure 7 also shows that very small amounts of pedogenic carbonate theoretically could accumulate at depths below 5 m,

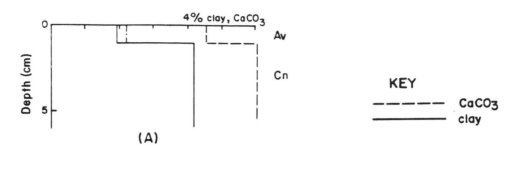

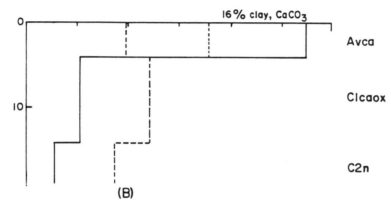

Figure 6. Carbonate and clay content (percent) for soils formed on Holocene deposits in an arid, hyperthermic climate in southern California (data from McFadden, 1982). Parent material is very gravelly lithic arkosic loamy sand. Soils associated with terrace deposits of Whipple Mountains in the eastern Mojave Desert. (A) Latest Holocene (<1,000 years B.P. (?)); (B) Middle Holocene (4,000 to 8,000 years B.P.).

well below the solum. Pedogenic carbonate, however, is not observed at these depths in most Holocene soils on the southern piedmonts of the Transverse Ranges. This lack of pedogenic carbonate is probably the result of high Li values (20 to 54 cm) associated with a moister mesic climate present near the range fronts (McFadden, 1982). Calcareous Holocene soils forming in the xeric, thermic climate of the eastern piedmont deposits of the central Coast Ranges of California studied by Lettis (1982; this volume) show that a drier variant of a Mediterranean climate is conducive to development of calcic soils.

PREDICTED IMPACT OF THE PLEISTOCENE-HOLOCENE CLIMATE CHANGE ON CALCIC SOIL DEVELOPMENT

Abundant paleobotanical, lacustrine, and geomorphological evidence demonstrates that the latest Pleistocene climate in the southwestern United States was substantially different than the Holocene climate (Flint, 1971; Wright and Porter, 1983. Many studies attribute the morphology of pedogenic carbonate observed in Pleistocene deposits to the effects of glacial-to-interglacial climatic changes (Gile and others, 1966; Gile and Grossman, 1979; Morrison and Frye, 1965; Nettleton and others,

1975; Shlemon, 1978; McFadden, 1981; McFadden and Bull, 1981). Birkeland (1974), Birkeland and Shroba (1974), and Shroba (1980) assert that much of the morphologic evidence often cited in support of polygenetic profile development can be related to causes other than climatic change, such as changes in the texture of parent material.

The final resolution of these differences in our opinions ultimately will require many additional field and laboratory soils analyses, precise ages of soil and parent material, and precise data regarding regional and temporal changes in the influx rates of carbonate (see Machette, this volume). The model we present can be used to evaluate whether or not climatically induced polygenesis in calcic soils is theoretically realistic.

The timing and relative magnitude of the Pleistocene-Holocene climatic change in the southwestern United States has been the subject of many studies and is becoming increasingly well defined (Smith, 1968; Van Devender, 1973, 1977; Bull, 1974; Spaulding, 1982; Wright and Porter, 1983). However, the type of climatic change required to recalculate the latest Pleistocene full-glacial soil-water balance for a given region is the subject of continuing debate. There are two principal schools of thought regarding the type of climatic change. A major decrease in precipitation coupled with an increase in mean annual temper-

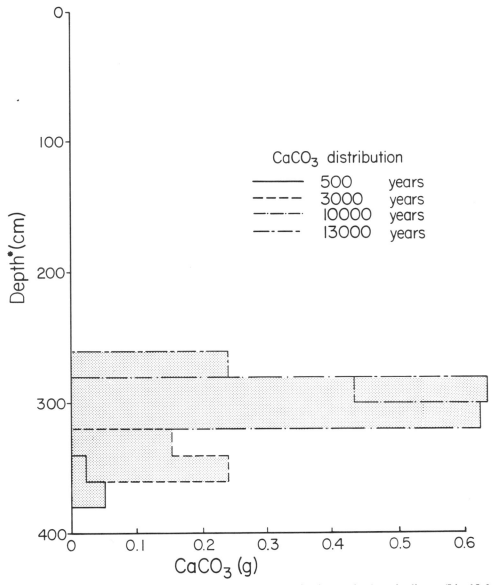

Figure 7. Predicted patterns of pedogenic carbonate accumulation in a xeric, thermic climate (Li = 18.6 cm; carbonate influx rate = 1.5×10^{-4} g/cm^2/yr; pCO$_2$ = $10^{-2.0}$ atm). No carbonate is present in the soil parent material. Gray area shows final distribution of carbonate. Depth* = absolute infiltration depth in the <2 mm size fraction.

ature of perhaps as little as 3°C has been proposed by Antevs (1952), Snyder and Langbein (1962), Flint (1971), Van Devender (1977), and Wells (1979). The other school proposes a major increase in annual temperature (>8°C) with little or no major changes in annual precipitation (Galloway, 1970; Brakenridge, 1978). Table 4 shows the recalculated late Pleistocene, full-glacial (pluvial) temperature, evapotranspiration, and leaching-index data for the presently arid, hyperthermic climate of the southeastern Mojave Desert. The data indicate that both postulated climates would be conducive to a major increase in effective soil moisture by a factor of 2.3 to 3.7. Using data from Tables 1, 2, 3, and 4, the effect of changing from calcic soil development in an interglacial or latest Pleistocene full-glacial climate to development in a Holocene climate can be predicted. Figure 9 shows the result of the predictions of the pattern of carbonate accumulation in late Pleistocene soils. The critical aspect of this figure is the bimodal distribution of carbonate in late Pleistocene soils. Part of the strong polygenetic bimodality is the result of the assumed abrupt climatic transition. Paleobotanical evidence (Van Devender, 1973, 1977) and pluvial lake chronologies (Smith, 1968; Smith, 1975), show that the early Holocene climate was transitional between the full-glacial and the warm, dry middle Holocene climate. For our model, this transitional climate change can be characterized by changing Li more gradually over the

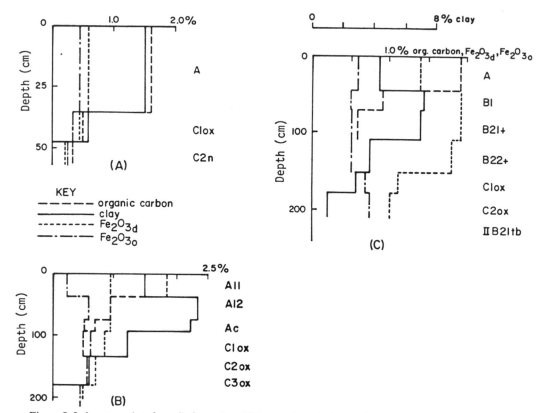

Figure 8. Laboratory data for soils formed on Holocene deposits in a xeric, thermic climate in southern California (data from McFadden, 1982). Parent material is very gravelly to bouldery lithic arkosic sand. Soils formed on terrace deposits of San Gorgonio Wash, central San Bernardino Mountains. (A) latest Holocene (<1,000 years B.P.). (B) late Holocene (1,000 to 3,000 years B.P.) (C) middle to early (?) Holocene 8,000 to 13,000 years B.P.).

TABLE 4. ESTIMATED EVAPOTRANSPIRATION (ETp) AND LEACHING INDEX (Li) FOR THE LATE PLEISTOCENE AND EARLY HOLOCENE CLIMATE OF THE PRESENTLY ARID, HYPERTHERMIC CLIMATE OF THE EASTERN MOJAVE DESERT, CALIFORNIA AND SOUTHEASTERN ARIZONA[1]

	J	F	M	A	M	J	J	A	S	O	N	D
							Month					
1. ET_p	0.28	1.01	$T^2 = -8°C$ $-$	$P^2 = 0$ 3.48	$-$	$-$	Li: 2.28 cm $-$	$-$	$-$	$-$	0.77	0.13
2. ET_p	0.39	1.16	$T = -6.5°C$ $-$	$P = 0$ $-$	$-$	$-$	Li: 1.85 cm $-$	$-$	$-$	$-$	$-$	0.56
3. Precipitation	1.68	1.25	$T = -5.0°C$ $-$	$P = +20\%$ $-$	$-$	$-$	Li: 2.02 cm $-$	$-$	$-$	$-$	2.28	1.68
ET_p	0.67	$-$	$-$	$-$	$-$	$-$	$-$	$-$	$-$	$-$	2.39	0.67
4. Precipitation	1.68	1.25	$T = -3.0°C$ $-$	$P = +20\%$ $-$	$-$	$-$	Li: 1.67 cm $-$	$-$	$-$	$-$	1.28	1.68
ET_p	0.68	1.52	$-$	$-$	$-$	$-$	$-$	$-$	$-$	$-$	2.39	1.01
5. Precipitation	2.10	1.55	$T = -3.0°C$ $-$	$P = +50\%$ $-$	$-$	$-$	Li: 2.45 cm $-$	$-$	$-$	$-$	1.60	2.10
ET_p	0.68	1.52	$-$	$-$	$-$	$-$	$-$	$-$	$-$	$-$	2.39	1.01
6. Precipitation	2.52	1.84	$T = -3.0°C$ $-$	$P = +80\%$ $-$	$-$	$-$	Li: 3.67 cm $-$	$-$	$-$	$-$	1.80	2.52
ET_p	0.68	1.52	$-$	$-$	$-$	$-$	$-$	$-$	$-$	$-$	2.39	1.01

[1]Pleistocene and early Holocene estimates calculated based on modern climatic data for Parker Reservoir, California (see Table 1 of this study), using different ΔT and ΔP combinations (1-6). Combinations 1 and 2 represent cold and dry climates based on data from Brakenridge(1978). Combinations 3-6 represent variants of cool and moist climates for latest Pleistocene and early Holocene based on data from Snyder and Langbein (1962), Flint (1971), Van Devender (1973,1977), and Wells (1979).

[2]Changes in T, P are considered to be uniform throughout year. Data are not shown for months when recalculated ET_p significantly exceeds recalculated precipitation.

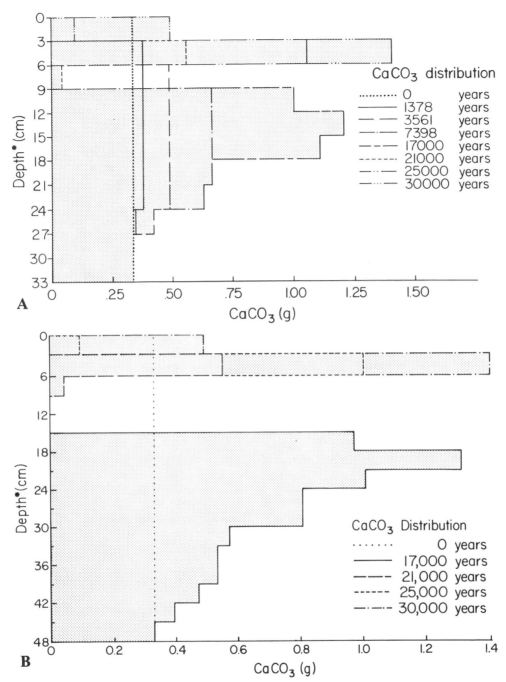

Figure 9. Predicted patterns of carbonate accumulation in a late Quaternary deposit (30,000 years old). Dotted line shows the carbonate distribution at the beginning of soil development (t_o) of 30,000 years B.P. The carbonate influx is 1.0×10^{-4} g/cm²/yr. (A) Pattern of carbonate accumulation assuming a change in climate from a relatively cold, dry late Pleistocene climate (30,000 to 17,000 years B.P.) to the Holocene climate (13,000 years B.P. to present). Under late Pleistocene climate, leaching index (Li) = 2.3 cm and $pCO_2 = 10^{-2.5}$ atm. Under Holocene climate, Li = 0.97 cm and $pCO_2 = 10^{-2.8}$ atm. Late Pleistocene to Holocene change in mean annual temperature = +8°C. No change in average monthly precipitation. Gray area shows distribution of carbonate after 30,000 years of soil development. (B) Pattern of carbonate accumulation assuming a change in climate from a relatively cool, moist late Pleistocene climate to the Holocene climate. Under late Pleistocene climate, Li = 2.3 cm and $pCO_2 = 10^{-2.5}$ atm. Holocene conditions are identical with those in 9A. The increase in average monthly precipitation, late Pleistocene to Holocene = +80 percent. Increase in mean monthly temperature, late Pleistocene to Holocene = +3°C. Gray area shows distribution of carbonate after 30,000 years of soil development. Depth* = absolute infiltration depth in the <2 mm size fraction.

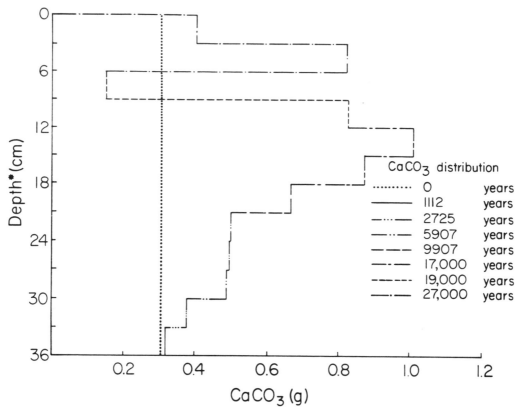

Figure 10. Predicted pattern of carbonate accumulation in a late Quaternary deposit assuming a Pleistocene/Holocene change from a cold, dry climate. Dotted line shows the carbonate distribution at the beginning of soil development 17,000 years ago (t_o). The carbonate influx is 1×10^{-4} g/cm^2/yr. Under the late Pleistocene climate, 27,000 to 14,000 years B.P., leaching index (Li) = 2.3 cm; pCO$_2$ from 0 to 9 cm = $10^{-2.5}$ atm; pCO$_2$ below 9 cm decreases to minimum of $10^{-3.4}$ atm in compartment 12 (33 to 36 cm). Under the latest Pleistocene/earliest Holocene climate, 14,000 to 12,000 years B.P., Li = 1.85 cm; pCO$_2$ characteristics are the same as those 27,000 to 14,000 years B.P. No change in mean monthly precipitation; 1.5°C increase in mean monthly temperature. Under early Holocene climate, 12,000 to 10,000 years B.P., Li = 2.02 cm, pCO$_2$ characteristics are unchanged. Change in mean monthly precipitation = +20 percent; 1.5°C increase in mean monthly temperature. Under early to middle Holocene climate, 10,000 to 8,000 years B.P., Li = 1.70 cm; pCO$_2$ characteristics are unchanged. Change in mean monthly precipitation = +20 percent; 3°C increase in mean monthly temperature. Under middle Holocene to present climate, 8,000 years B.P. to present, Li = 0.97 cm; pCO$_2$ = $10^{-2.8}$ atm in uppermost three compartments decreasing below third compartment to a minimum of $10^{-3.4}$ atm. Change in mean monthly precipitation = –40 percent; 2°C increase in mean monthly temperature. Different line patterns show the time period during which maximum carbonate distribution in a given compartment is attained. Depth* = absolute infiltration depth in the <2 mm size fraction.

period 13,000 to 8,000 years B.P. Even considering this modification (shown in Fig.10), there is still a strongly bimodal distribution.

Figure 11 shows profile data for soil typical of the late Pleistocene Q2b geomorphic surfaces defined by Bull (1974). Bull has assigned an age of 11,000 to 70,000 years B.P. to this surface based on geomorphological evidence and ^{230}Th/^{234}U dates on carbonate. The whole-profile index of secondary carbonate (cS, Machette, this volume) in these soils (3.8 and 6.6 g/cm^2-soil column) supports Bull's age estimates, as discussed in McFadden (1982). The predicted and actual patterns of carbonate distribution for late Pleistocene soil profiles shown in Figures

9, 10, and 11 are similar; this similarity indicates theoretical support for the genesis of polygenetic calcic soils in arid and semiarid climates.

Our model predicts that a polygenetic profile could form as a result of a change from a cold glacial climate to a much warmer interglacial climate. No associated decrease in precipitation is required; however, much smaller increases in mean monthly temperature in the early Holocene would require concomitant increases in precipitation in order to achieve a polygenetic soil profile. It is also likely that a wet pluvial climate would not be conducive to polygenetic calcic soil development if increases in precipitation were confined primarily to summer months.

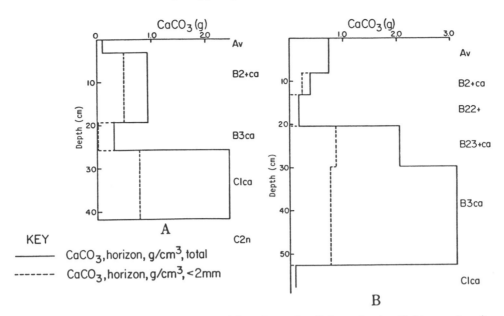

Figure 11. Carbonate and clay content (percent) for polygenetic soils formed on late Pleistocene deposits in the arid, hyperthermic eastern Mojave Desert of southern California (data from McFadden, 1982). Parent material is very gravelly lithic arkosic loamy sand. Soils associated with terrace deposits of Whipple Mountains. (A) Soil on latest Pleistocene of youngest recognized phase of the Q2c geomorphic surface. (B) Soil on latest Pleistocene of oldest recognized phase of the Q2c geomorphic surface.

Summer evapotranspiration rates would not be low enough and carbonate solubility would not be increased enough during the winter so as to increase Li sufficiently (see Tables 2 and 4) to cause the locus of carbonate accumulation to be lowered markedly relative to levels observed in Holocene soil profiles. The predictions also show that the development of an increasingly finer soil texture due to clay accumulating in the B horizon could not cause the massive B horizon to be engulfed by carbonate (see Fig. 11). Indeed, a concomitant effect of depleting rather than accumulating carbonate from the upper part of the soil profile during the moister late Pleistocene climate may have been to accelerate the development of the argillic B horizon due to enhanced dispersal and translocation of clay in an essentially noncalcareous zone above the calcic horizon. Prominent argillic B horizons of soils on late Pleistocene deposits in arid, hot deserts thus may reflect increased illuviation of eolian clay as well as the possibly greater magnitude and rate of chemical weathering which prevailed during glacial-age climates.

DISCUSSION

Previous studies of calcic soil development have emphasized empirical approaches. The compartment model for predicting patterns of carbonate accumulation provides a more theoretical approach to the study of calcic-soil development. Application of the model to many Quaternary studies appears to be particularly promising. The availability of data such as detailed soil pCO_2 measurements and carbonate influx rates for different areas during the Quaternary undoubtedly will increase the accuracy of and confidence in predicted characteristics of profiles. For example, many problems such as the occurrence of apparently nonpolygenetic Pleistocene calcic and noncalcic soils in regions where the climate presently is arid to semiarid, the often complex and enigmatic nature of the so-called Pedocal-Pedalfer boundary, or the possible impact of an interglacial climate on pre-Wisconsinan calcic soils, can be evaluated more rigorously by integrating theoretical models with field and laboratory soil research.

ACKNOWLEDGMENTS

Our sincere thanks to the U.S. Geological Survey which directly or indirectly supported most of the research upon which this study is based. Dr. William B. Bull has been a major source of inspiration with respect to many of our ideas concerning paleoclimates and desert soils. We thank Austin Long for important discussions and critiques of early drafts of this paper. We also thank Brian Atwater and especially Michael N. Machette for their reviews of later drafts of this paper. We appreciate useful comments of Dan H. Yaalon, C. Vance Haynes, G. Robert Brakenridge, Peter S. Homann, and Rodney J. Arkley. The idea of using a compartment-model strategy was inspired by Robert J. Rogers, to whom the senior author is especially indebted. Thanks also to Judy A. Salas for drafting and Amy A. Tokunaga for patience in typing the manuscripts. Dorothy J. Merritts assisted in the preparation of the final manuscript. This study was presented at the 78th Annual GSA Cordilleran Section Meeting in Anaheim, California, in April, 1982, at a symposium dedicated to the late Denis E. Marchand.

REFERENCES CITED

Antevs, E., 1952, Cenozoic climates of the Great Basin: Geologische Rundschau, v. 40, p. 94–108.

Arkley, R. J., 1962a, The geology, geomorphology, and soils of the San Joaquin Valley in the vicinity of the Merced River, California, *in* Geologic guide to the Merced Canyon and Yosemite Valley: California Division of Mines and Geology, Bulletin 182, p. 25–32.

—— 1962b, Soil survey of the Merced area, California: U.S. Department of Agriculture, Soil Survey Series, 1950, no. 7, 131 p.

—— 1963, Calculations of carbonate and water movement in soil from climatic data: Soil Science, v. 96, no. 4, p. 239–248.

—— 1964, Soil survey of the eastern Stanislaus area, California: U.S. Department of Agriculture, Soil Survey Series 1957, no. 20, 160 p.

Bachman, G. O., and Machette, M. N., 1977, Calcic soils and calcretes in the southwestern United States: U.S. Geological Survey Open-File Report 77-794, 163 p.

Barnes, I., 1965, Geochemistry of Birch Creek, Inyo County, California, a travertine-depositing creek in an arid climate: Geochimica et Cosmochimica Acta, v. 29, p. 85–112.

Baver, L. D., 1956, Soil Physics: New York, John Wiley and Sons, 489 p.

Birkeland, P. W., 1974, Pedology, weathering, and geomorphological research: New York, Oxford University Press, 285 p.

Birkeland, P. W., and Shroba, R. R., 1974, The status of the concept of Quaternary soil-forming intervals in the western United States, *in* Mahany, W. C., ed., Quaternary Environments: Symposium Proceedings, Geological Monographs, York University, Toronto, p. 241–276.

Blatt, H., Middleton, G. V., and Murray, R. C., 1981, Origin of sedimentary rocks: Englewood Cliffs, New Jersey, Prentice-Hall, Inc., 782 p.

Bogli, A., 1980, Karst hydrology and physical speleology: New York, Springer-Verlag, 328 p.

Bohn, H. L., McNeal, B. L., and O'Conner, G. A., 1979, Soil Chemistry: New York, John Wiley and Sons, 329 p.

Brakenridge, G. R., 1978, Evidence for a cold, dry full-glacial climate in the American Southwest: Quaternary Research, v. 9, p. 22–40.

Bull, W. B., 1974, Geomorphic tectonic analysis of the Vidal region, *in* Information concerning site characteristics, Vidal Nuclear Generating Station [California]: Los Angeles, Southern California Edison Company, Appendix 2.5B, amendment 1, 66 p.

Busenberg, E., and Clemency, C. V., 1976, The dissolution kinetics of feldspars at 25° and 1 atm CO_2 partial pressure: Geochimica et Cosmochimica Acta, v. 40, p. 41–49.

Drever, J. I., 1982, The geochemistry of natural waters: Englewood Cliffs, New Jersey, Prentice-Hall, Inc., 388 p.

Dreybrodt, W., 1981, Kinetics of the dissolution of calcite and its applications to karstification: Chemical Geology, v. 31, p. 245–269.

Flint, R. F., 1971, Glacial and Quaternary geology: New York, John Wiley and Sons, 892 p.

Frissel, M. J., and Reineger, P., 1974, Simulation of accumulation and leaching in soils: Wageninger, Netherlands, Centre for Agricultural Publications and Documents, 116 p.

Galloway, R. W., 1970, The full-glacial climate in the southwestern United States: Annals of the Association of American Geographers, v. 60, p. 245–256.

Gardner, R. L., 1972, Origin of the Mormon Mesa Caliche, Clark County, Nevada: Geological Society of America Bulletin, v. 83, p. 143–155.

Garrels, R. M., and Christ, C. L., 1965, Solutions, minerals, and equilibria: New York, Harper and Row, 450 p.

Garrels, R. M., and Mackenzie, F. T., 1971, Evolution of sedimentary rocks: New York, W. W. Norton and Company, Inc., 397 p.

Gile, L. H., 1975, Holocene soils and soil-geomorphic relations in an arid region of southern New Mexico: Quaternary Research, v. 5, p. 321–360.

—— 1977, Holocene soils and soil-geomorphic relations in a semiarid region of southern New Mexico: Quaternary Research, v. 7, p. 112–132.

Gile, L. H., and Grossman, R. B., 1979, The desert soil project monograph: U.S. Department of Agriculture, Soil Conservation Service, 984 p.

Gile, L. H., Peterson, F. F., and Grossman, R. B., 1966, Morphological and genetic sequences of carbonate accumulation in desert soils: Soil Science, v. 101, p. 347–360.

Griffin, R. A., and Jurinak, J. J., 1973, Estimation of activity coefficients from an electrical conductivity of natural aquatic system and soil extracts: Soil Science, v. 116, p. 26–30.

Harden, J. W., and Marchand, D. E., 1977, The soil chronosequence of the Merced River area [California], *in* Singer, M. J., ed., Soil development, geomorphology, and Cenozoic history of the northeastern San Joaquin Valley and adjacent area, California: Joint Field Session, American Society for Agronomy, Soil Science Society of America, and Geological Society of America, Guidebook, Davis, University of California Press, p. 22–38.

Hopkins, D. M., 1975, Time-stratigraphic nomenclature for the Holocene Epoch: Geology, v. 3, p. 10.

Jacobson, L. J., and Langmuir, D. M., 1974, Dissociation constant of calcite and $CaHCO_3^-$ from 0° to 50°C: Geochimica et Cosmochimica Acta, v. 38, p. 301–318.

Janda, R. J., and Croft, M. G., 1967, The stratigraphic significance of a sequence of noncalcic brown soils formed on Quaternary alluvium of the northeastern San Joaquin Valley, California, *in* Morrison, R. B., and Wright, H. E., Jr., eds., Quaternary Soils: Proceedings, International Association for Quaternary Research (INQUA), VII Congress, Reno, Nevada, v. 9 (Desert Research Institute), p. 157–190.

Jenne, E. A., 1971, Controls on Mn, Fe, Co, Ni, Cu, and Zn concentrations in soils and water: The significant role of hydrous Mn and Fe oxides, *in* Gould, R. F., ed., Nonequilibrium systems in natural water chemistry: American Chemical Society Advanced Chemistry Series, no. 106, p. 337–387.

Jenny, H., 1941, Factors of soil formation: New York, McGraw-Hill Book Company, 281 p.

—— 1980, The soil resource, origin and behavior: New York, Springer-Verlag, Ecological Studies 37, 377 p.

Junge, C. E., and Werby, R. T., 1958, The concentration of chloride, sodium, potassium, calcium, and sulfate in rainwater over the United States: Journal of Meterology, v. 15, p. 417–425.

Kline, J. R., 1973, Mathematical simulation of soil-plant relationships and soil genesis: Soil Science, v. 115, p. 240–249.

Krauskopf, K. B., 1967, Introduction to geochemistry: New York, McGraw-Hill Book Company, 721 p.

Lettis, W. R., 1982, Late Cenozoic stratigraphy and structure of the western margin of the central San Joaquin Valley, California: U.S. Geological Survey Open-File Report 82-526, 203 p.

—— 1985, Late Cenozoic stratigraphy and structure of the west margin of the central San Joaquin Valley, California, *in* Weide, D. L., ed., Soils and Quaternary geology of the southwestern United States: Geological Society of America Special Paper 203 (this volume).

Lindsay, W. L., 1979, Chemical equilibria in soils: New York, John Wiley and Sons, 329 p.

Machette, M. N., 1978, Dating Quaternary faults in the southwestern United States by using buried calcic paleosols: U.S. Geological Survey Journal of Research, v. 6, p. 369–381.

—— 1985, Calcic soils of the American Southwest, *in* Weide, D. L., ed., Soils and Quaternary geology of the southwestern United States: Geological Society of America Special Paper 203 (this volume).

Marchand, D. E., and Allwardt, A., 1981, Late Cenozoic stratigraphic units, northeastern San Joaquin Valley, California: U.S. Geological Survey Bulletin 1470, 70 p.

McFadden, L. D., 1981, Geomorphic processes influencing the Cenozoic evolution of the Canada del Oro Valley, southern Arizona, *in* Stone, C., and Jenney, J. P., eds.: Arizona Geological Society Digest, v. 8, p. 13–20.

—— 1982, The impacts of temporal and spatial climatic changes on alluvial soils genesis in southern California [Ph.D. thesis]: Tucson, University of Arizona, 430 p.

McFadden, L. D., and Bull, W. B., 1981, Impact of Pleistocene-Holocene climatic change on soils genesis in the eastern Mojave Desert, California [abs.]: Geological Society of America Abstracts with Programs, v. 13, p. 95.

McFadden, L. D., Tinsley, J. C., and Hendricks, D. M., 1980, A preliminary soil chronosequence for the tectonically and climatically controlled terraces and alluvial fans of the San Gabriel Mountains and the Los Angeles Basin, California [abs.]: Geological Society of America Abstracts with Programs, v. 12, p. 119.

McHargue, L. E., 1981, Late Quaternary deposition and pedogenesis on the Aguila Mountains piedmont, southwestern Arizona [M.S. thesis]: Tucson, University of Arizona, 132 p.

Morrison, R. B., and Frye, J. C., 1965, Correlation of middle and late Quaternary successions of Lake Lahontan, Lake Bonneville, Rocky Mountain (Wasatch Range), southern Great Plains and eastern Midwest areas: Nevada Bureau of Mines Report 9, 45 p.

National Oceanic and Atmospheric Administration, 1978, Climatologic Data, Annual Summary: Environmental Data and Information Service, U.S. Department of Commerce, v. 82, no. 13, 570 p.

Nettleton, W. D., Witty, J. E., Nelson, R. E., and Hawley, J. W., 1975, Genesis of argillic horizons in soils of desert areas of the southwestern United States, *in* Proceedings, Soil Science Society of America, v. 39, p. 919–926.

Palmer, W. C., and Havens, A. V., 1958, A graphical technique for determining evapotranspiration by the Thornthwaite method: Monthly Weather Review, April, p. 123–128.

Parada, C. B., Long, A., and Davis, S. N., 1983, Stable isotope composition of soil carbon dioxide in the Tucson Basin, Arizona: Isotope Geoscience, v. 1, p. 219–236.

Petraitus, M. J., and Wood, W. W., 1982, Partial pressure of CO_2, O_2, N_2, and Ar in a thick unsaturated zone of a semi-arid area [abs.]: Geological Society of America Abstracts with Programs, v. 14, p. 586.

Picknett, R. G., Bray, L. G., and Stenner, R. D., 1976, The chemistry of cave water, *in* Ford, D. T., and Cullingford, H. D., eds., The Science of speleology: London, Academic Press, p. 213–262.

Plummer, L. N., and Wigley, T.M.L., 1976, The dissolution of calcite in CO_2 saturated solution at 25°C and 1 atmosphere total pressure: Geochimica et Cosmochimica Acta, v. 40, p. 191–202.

Plummer, L. N., Wigley, T.M.L., and Parkhurst, D. L., 1978, The kinetics of calcite dissolution in CO_2-water systems at 5 to 60°C and 0.0 to 1.0 atm CO_2: American Journal of Science, v. 278, p. 179–216.

Reddy, M. M., and Nancollas, G. H., 1971, The crystallization of calcium carbonate: I. Isotopic exchange and kinetics: Journal of Colloid Interface, v. 36, p. 166–172.

Rogers, R. J., 1980, A numerical model for simulating pedogenesis in semiarid regions [Ph.D. thesis]: Salt Lake City, University of Utah, 285 p.

Schenker, A. R., 1978, Particle-size distribution of late Cenozoic gravels on an arid region piedmont, Gila Mountains, Arizona [M.S. thesis]: Tucson, University of Arizona, 118 p.

Sehgal, J. L., and Stoops, G., 1972, Pedogenic calcite accumulation in arid and semiarid regions of the Indo-Gangetic alluvial plain of erstwhile Punjab [India]: Their morphology and origin: Geoderma, v. 8, p. 59–72.

Shlemon, R. J., 1978, Buried calcic paleosols near Riverside, California, *in* Geologic guidebook to the Santa Ana River basin, southern California: South Coast Geological Society [Irvine, California], Annual Field Trip Guidebook, p. 32–40.

Shroba, R. R., 1980, Influence of parent material, climate, and time on soils formed in Bonneville-shoreline and younger deposits near Salt Lake City and Ogden, Utah [abs.]: Geological Society of America Abstracts with Programs, v. 12, p. 304.

Siever, R., and Woodford, N., 1979, Dissolution kinetics and the weathering of mafic minerals: Geochimica et Cosmochimica Acta, v. 43, p. 717–724.

Smith, G. I., 1968, Late Quaternary geologic and climatic history of Searles Lake, southeastern California, *in* Morrison, R. B., and Wright, H. E., Jr., eds., Means of correlation of Quaternary successions: Proceedings, International Association for Quaternary Research (INQUA), VII Congress, Salt Lake City, Utah, V. 8 (University of Utah Press), p. 293–310.

Smith, R.S.U., 1975, Late Quaternary pluvial and tectonic history of Panamint Valley, Inyo and San Bernardino Counties, California [Ph.D. thesis]: Pasadena, California Institute of Technology, 295 p.

Snyder, C. T., and Langbein, W. B., 1962, The Pleistocene lake in Spring Valley, Nevada, and its climatic implications: Journal of Geophysical Research, v. 67, p. 2385–2394.

Sohl, N. F., and Wright, W. B., 1980, Changes in stratigraphic nomenclature by the U.S. Geological Survey, 1979: U.S. Geological Survey Bulletin 1502-A, p. A1–A2.

Soil Survey Staff, 1975, Soil taxonomy: A basic system of soil classification for making and interpreting soil surveys: U.S. Department of Agriculture, Soil Conservation Service, Agricultural Handbook No. 436, Washington, D.C., U.S. Government Printing Office, 754 p.

Spaulding, W. G., 1982, Processes and rates of vegetation change in the arid southwest [abs.]: American Quaternary Association, Program and Abstracts, Seventh Biennial Conference, p. 22–24.

Suarez, D. L., 1981, Predicting Ca and Mg concentrations in arid land soils: EOS (American Geophysical Union Transactions), v. 62, p. 285.

Suarez, D. L., and Rhoades, J. D., 1982, The apparent solubility of calcium carbonate in soils: Soil Science, v. 134, p. 716–722.

Tinsley, J. C., DesMarais, D. J., McCoy, G., Rogers, B. W., and Ulfeldt, S. R., 1981, Lilburn Cave's contributions to the natural history of Sequoia and Kings Canyon National Parks, California, USA, *in* Proceedings, Eighth International Congress of Speleology, v. 1, p. 287–289.

Van Devender, T. R., 1973, Late Pleistocene plants and animals of the Sonoran Desert: A survey of ancient packrat middens in southwestern Arizona [Ph.D. thesis]: Tucson, University of Arizona, 179 p.

——1977, Holocene woodlands in the southwestern deserts: Science, v. 198, p. 189–192.

Wells, P. V., 1979, An equable glaciopluvial in the west: Pleniglacial evidence of increased precipitation on a gradient from the Great Basin to the Sonoran and Chihuahuan Deserts: Quaternary Research, v. 12, p. 311–325.

Wright, H. E., Jr., and Porter, S. C., 1983, Late-Quaternary environments of the United States: Minneapolis, University of Minnesota Press, 683 p.

MANUSCRIPT ACCEPTED BY THE SOCIETY JANUARY 12, 1985

Geological Society of America
Special Paper 203
1985

Quaternary deposits and soils in and around Spanish Valley, Utah

Deborah R. Harden
Norma E. Biggar
Mary L. Gillam
Woodward-Clyde Consultants
1 Walnut Creek Center
100 Pringle Avenue
Walnut Creek, California 94596

ABSTRACT

Examination of Quaternary deposits in Spanish Valley near Moab, Utah, revealed a complex pattern of soil development and distribution on alluvial fans and terraces previously correlated by Richmond (1962) with glacial deposits in the La Sal Mountains. Longitudinal profiles show downstream convergence of terraces and burial of older surfaces in the middle portion of the valley. In this area, buried soils at two localities have carbonate accumulations similar to those on Placer Creek terraces considered to be of Bull Lake age. The reappearance of inset terraces in lower Spanish Valley suggests that the depositional history has been complex or that the middle part of the valley has subsided since Placer Creek time.

Progressive changes in the morphology and content of pedogenic carbonate were used to correlate deposits. Soils developed on gravel of the latest Pleistocene Beaver Basin Formation (Pinedale age) have thin carbonate coats on clasts, display Stages I and II carbonate morphology, and contain 7 to 9 gm/cm^2 of pedogenic CaCO$_3$. Placer Creek deposits have up to 75 cm of continuous CaCO$_3$; three sampled profiles contain about 36 to 38 gm/cm^2. Carbonate contents of soils on the middle Pleistocene Middle Member of the Harpole Mesa Formation are similar to those of Placer Creek soils; however, a buried argillic horizon preserved at one locality suggests that the Harpole Mesa deposit is older. The Stage IV K horizon on older Harpole Mesa alluvium has a strong platy structure and a well-developed laminar horizon; about 182 gm/cm^2 of pedogenic CaCO$_3$ is present in one profile. Sandy alluvium in the Lower Member of the Harpole Mesa Formation has reversed magnetic polarity. A similar pedogenic calcrete containing 98 to 130 gm/cm^2 of accumulated CaCO$_3$ overlies magnetically reversed deposits on an erosion surface southwest of Green River, Utah. Using a minimum age of 700,000 years for the magnetically reversed sediments, maximum carbonate accumulation rates for this portion of Utah range from 0.14 to 0.26 gm/cm^2/1,000 years.

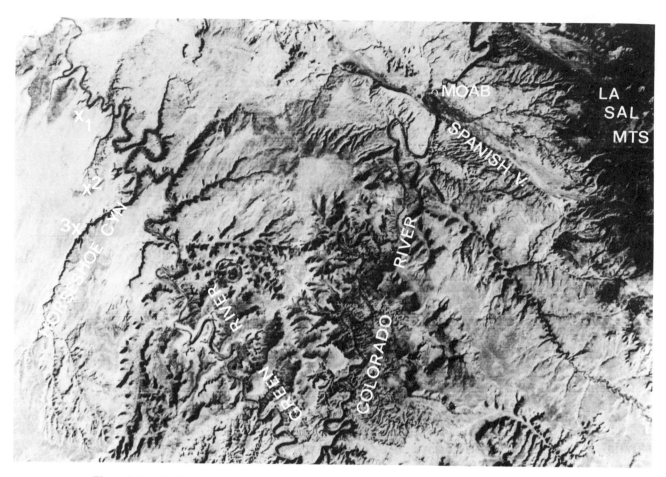

Figure 1. Landsat image showing location of the study area in southeastern Utah. Approximate scale 1:500,000. Numbered localities refer to soil profiles shown in Figure 15.

INTRODUCTION

Quaternary deposits in southeastern Utah record several interruptions of the long-term denudation that has produced the striking erosional landforms characteristic of the Canyonlands area. In Spanish Valley, located on the western and southwestern flanks of the La Sal Mountains (Fig. 1), Quaternary alluvium ranging in age from modern to early Pleistocene is preserved in valley fills and terraces of Pack Creek and Mill Creek.

The topographic position of terraces and the relative development of calcic soils, together with limited radiocarbon dates and paleomagnetic data, provide a means of correlating the deposits and establishing a soil chronosequence for the area. Deposits and soils of probable early Pleistocene age also occur west of the Green River (Fig. 1) and can be correlated with those in Spanish Valley.

SCOPE OF WORK

Our work in Spanish Valley is part of a regional study of Quaternary deposits, ground water, geomorphology, and soils in the Paradox basin, southeastern Utah, to assess the tectonic stabil-

ity, erosion rates, and climatic history of the area. Because the Quaternary stratigraphy of the contiguous La Sal Mountains and vicinity had previously been established by Richmond (1962), we focused on that area to provide a framework for the Paradox basin as a whole. Spanish Valley has a climatic setting and terrace deposits similar to those in much of southeastern Utah. The study sites in Spanish Valley were located on successively older alluvial deposits underlying vertically distinct terrace remnants.

PHYSIOGRAPHIC AND CLIMATIC SETTING

The climate of most of Spanish Valley is arid, with semiarid conditions prevailing at elevations above about 2,000 m. The average annual precipitation at Moab for the period 1900 through 1968 is 208 mm, and the mean annual precipitation for Spanish Valley (Fig. 1) is 381 mm (Sumsion, 1971). The higher portions of the La Sal Mountains receive more than 762 mm of precipitation annually. Much of the winter precipitation is snow, and the summer precipitation falls during local, intense thunderstorms. The mean annual temperature at Moab is 13.3°C (Sumsion, 1971).

The vegetation distribution in Spanish Valley is mainly in-

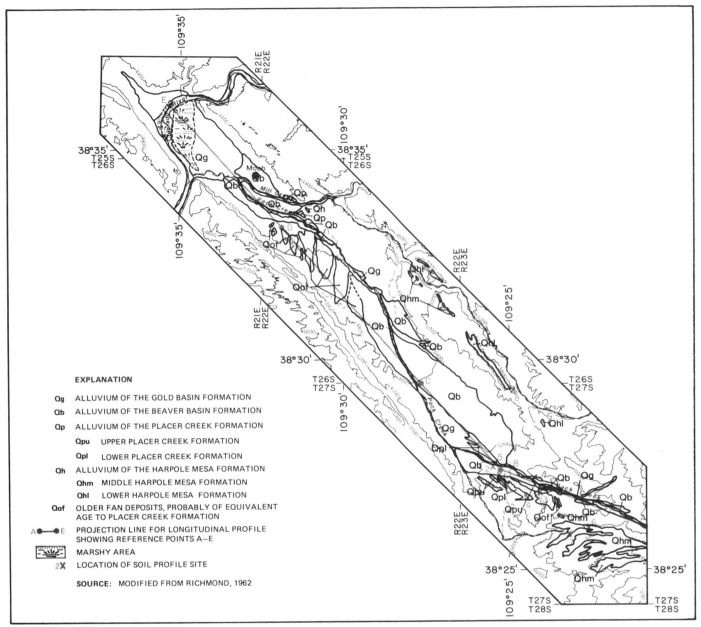

Figure 2. Map showing Quaternary deposits in Spanish Valley, modified from Richmond (1962). Base map from U.S. Geological Survey 1:24,000 scale topographic maps.

fluenced by available moisture and correlates closely with elevation. Below about 2,135 m sagebrush and grass are dominant. Pinyon-juniper and scrub oak-mountain mahogany forests occur between about 1,675 and 2,900 m. Discontinuous areas between 2,440 and 3,500 m are characterized by subalpine forest, and the upper tree line is between 3,350 and 3,500 m (Iorns and others, 1965).

Spanish Valley is drained by Pack Creek, which flows about 35 km from its headwaters in the La Sal Mountains to its junction with Mill Creek in the town of Moab (Fig. 2). Mill Creek, which also drains the La Sal Mountains, joins the Colorado River about 3 km west of Moab. Both Pack Creek and Mill Creek are natu-

rally perennial streams, but both streams are presently diverted for irrigation in the middle portion of Spanish Valley. Their combined drainage area is about 373 km² (Sumsion, 1971).

GEOLOGIC SETTING

Intrusive igneous rock of Tertiary age underlies the highest portions of the Pack Creek and Mill Creek drainages in the La Sal Mountains (Hunt, 1956). The composition of the intrusive rocks ranges from diorite to syenite. The Mancos and Morrison Formations of Cretaceous and Jurassic age, respectively, are exposed along the margins of the La Sal Mountains. Resistant sandstones of the Jurassic and Triassic Navajo, Kayenta, and Wingate Formations are exposed on the mesas and hogbacks flanking Spanish Valley (Williams, 1964). Exposures of Paleozoic rock are found only on the sides of the valley near the mouth of Mill Creek, where strata of the Cutler Group crop out.

Spanish Valley is a northwest-trending, fault-bounded trough on the west flank of the La Sal Mountains (Figs. 1 and 2). It is one of several salt anticlinal valleys in southeastern Utah and southwestern Colorado (Hunt, 1956). Subsurface information indicates that the salt anticlines are diapiric structures formed in the Pennsylvanian Paradox Member of the Hermosa Formation. In Spanish Valley, the diapirs pierce the overlying brittle strata, and the uppermost part of the structure has been removed by erosion. Subsurface well data indicate that 2,750 to 3,650 m of salt, overlain by up to 300 m of Hermosa cap rock (solution residue) and as much as 600 m of collapsed post-Hermosa strata, underlie the floor of Spanish Valley. The axis of Spanish Valley is roughly coincident with the Moab anticline in the northwestern part of the valley and with the Spanish Valley-Pack Creek syncline at the southeastern end (Williams, 1964). The folds probably formed in response to dissolution and diapiric migration of salt at depth. Several discontinuous normal faults mapped along the valley borders reflect collapse of the overlying strata following near-surface removal of salt in the valley, and possibly dissolution and (or) flowage of salt at depth.

PREVIOUS WORK

Quaternary deposits in the Spanish Valley area were studied in detail by Richmond (1962), who correlated moraines, outwash terraces, landslides, and eolian deposits with multiple glaciations in the La Sal Mountains. On the basis of relative topographic position, lithology and texture of deposits, and degree of soil development, Richmond grouped the deposits into the Gold Basin, Beaver Basin, Placer Creek, and Harpole Mesa Formations (Table 1), each of which was subdivided into members and correlated with recognized glacial-interglacial cycles in the Rocky Mountain region. Richmond realized that many of the surfaces in the area are underlain by complex deposits and soils of different ages and origins.

Recent work in the La Sal Mountains by Shroder and others (1980) suggests that glacial deposits are much less extensive than is indicated by Richmond's 1962 map. According to Shroder and others (1980), widespread diamictons which Richmond previously mapped as glacial deposits are actually mass movement deposits

QUATERNARY DEPOSITS

Quaternary deposits in Spanish Valley include alluvium, eolian deposits, and colluvium. Locally derived talus, landslide debris, and deposits of small fans occur along the margins of Spanish Valley, and recent eolian deposits are common throughout the area. The alluvium deposited by Pack Creek and Mill

Creek, which is the focus of this study, records Quaternary climatic fluctuations in the La Sal Mountains according to Richmond (1962). Quaternary alluvial deposits underlie terraces and pediments adjacent to Pack Creek and Mill Creek (Figs. 2 and 3). The deposits typically consist of poorly sorted, crudely stratified, sandy cobble gravel. Boulders up to 1.2 m in diameter are also present in the deposits. Clasts of porphyritic intrusive rock, derived from the La Sal Mountains, constitute the dominant lithology; however, sandstone clasts are also common.

Alluvium of the Lower Member of the Harpole Mesa Formation of early Pleistocene age (Richmond, 1962) occurs on the northeastern rim of Spanish Valley, and deposits identified as the Middle Member of the Harpole Mesa Formation underlie a south-dipping pediment along the southern margin of the valley (Figs. 2 and 3). Upper and middle Pleistocene deposits of the Placer Creek and Beaver Basin Formations occur as nested terraces in upper Spanish Valley and, downstream, below the junction of Pack and Mill Creeks (Fig. 2). Water-well logs (Sumsion, 1971, Utah Division of Water Rights, unpublished data) show that as much as 85 m of gravel is present beneath the Gold Basin and Beaver Basin surfaces in middle Spanish Valley. However, gravel deposits underlying the Beaver Basin Formation near the confluence of Pack and Mill Creeks are less than 47 m thick.

Fine-grained alluvium and eolian material are characteristic of Holocene deposits in Spanish Valley. Fine-grained alluvium of the Gold Basin Formation occurs as low terrace deposits along Pack Creek and its tributaries in the northwestern part of Spanish Valley (Fig. 2). The finer grain size distinguishes the Gold Basin Formation from older alluvium; gravel lenses are only locally present. However, gravel is present in the active channels of Pack Creek and Mill Creek. Up to 38 cm of unstratified sand and silt overlie the gravelly alluvium on all Pleistocene terraces examined. These deposits are probably Gold Basin or Beaver Basin in age and eolian in origin. Alternatively, they could represent the uppermost deposits of upward-fining alluvial sequences.

TERRACE PROFILES

Longitudinal profiles of terrace remnants (Fig. 4), together with the presence of buried calcic soils in the middle portion of Spanish Valley, suggest the possible concurrent influence of structure and hydrologic variables on stream behavior and on the deposition of fluvial deposits. Reconstruction of paleostream gradients is hampered by the scarcity of terraces in the middle portion of the valley; however, the profiles suggest a general decrease in terrace gradients with time and a convergence of surfaces in the middle of Spanish Valley (Fig. 4). The convergence of terraces is indicative of a progressive decrease in stream gradient, which could have resulted from a decrease in the discharge of Pack Creek during Quaternary time.

The distribution of Quaternary alluvial deposits in Spanish Valley suggests that Pack Creek has been confined within Spanish Valley since deposition of the Harpole Mesa Formation. The surfaces underlain by the middle and lower members of the Har-

TABLE 1. CORRELATION OF QUATERNARY DEPOSITS IN SPANISH VALLEY
AND WEST OF THE GREEN RIVER

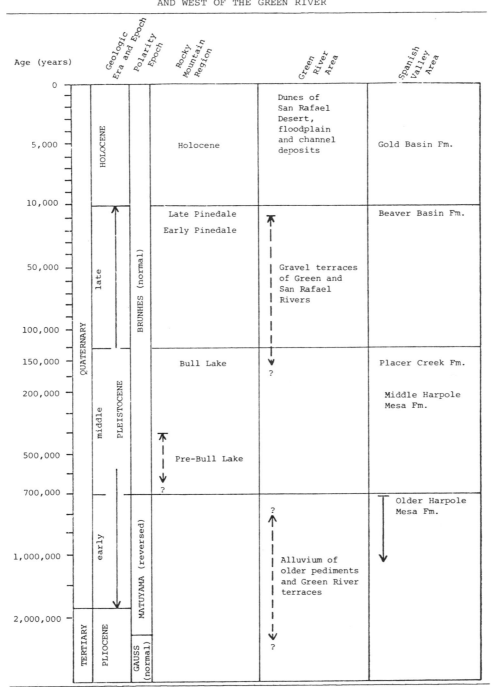

Note: Chronology modified from Meierding and Birkeland (1980), chronology
of Bull Lake glaciation from Madole and Shroba (1979), and obsidian hydration dating (Pierce and others, 1976) suggest that the Bull Lake glaciation correlates with a worldwide climatic event 130,000 to 150,000 years
B.P. that is recorded in deep sea cores (Shackleton and Opdyke, 1973). The
correlation of the La Sal Mountain deposits with Rocky Mountain glacial
deposits was made on the basis of Richmond's (1962) chronology and on the
basis of relative soil profile development on Spanish Valley Terraces.

Figure 3. View east toward La Sal Mountains showing Quaternary surfaces in Upper Spanish Valley.
GB: Gold Basin; BB: Beaver Basin; PC: Placer Creek; MHM: Middle Harpole Mesa.

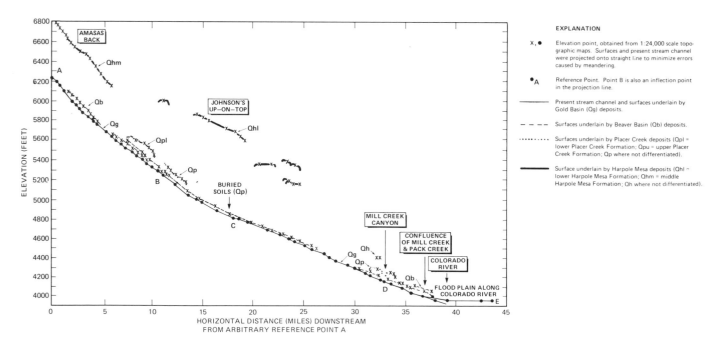

Figure 4. Longitudinal profiles of Pack Creek and terraces in Spanish Valley. The project line and
surfaces are shown in Figure 2. Identification and correlation of surfaces are based on Richmond (1962).

pole Mesa Formation along the northeastern and southern rims of the valley (Fig. 2) developed prior to formation of Spanish Valley. The shapes of the alluvial fans and terraces of Placer Creek and younger age reflect structural control of Spanish Valley by the bedrock faults that bound it. The shapes of terrace remnants in the upper portion of Spanish Valley (Fig. 2) suggest that terraces of a through-flowing single channel and of alluvial fans are both present there. The map pattern of Quaternary units also shows that the apex of each successively younger fan is at a greater distance from the present mountain front. Alluviation in the valley also occurs at the downstream end of the Pack Creek canyon, where the stream channel abruptly widens at the margin of Spanish Valley (Fig. 2).

Harpole Mesa and Placer Creek terrace remnants are absent in the middle portion of Spanish Valley. Richmond (1962) inferred the presence of a Placer Creek terrace surface beneath present stream level. The inferred buried surface occurs between Placer Creek terrace remnants that are tens of meters above present stream level at the mouth of Mill Creek and those in upper Spanish Valley (Figs. 2 and 4). Exposures created since Richmond's (1962) study support this inferred complexity: field observations at two excavations in the middle portion of the valley (Fig. 4) revealed two buried calcic soils that have carbonate morphology and content similar to that observed on terraces mapped as Placer Creek Formation and Middle Harpole Mesa Formation (Fig. 5). Richmond (1962) suggested that up to 23 m of erosion of the Placer Creek deposits occurred in the upper part of the valley prior to deposition of the Beaver Basin deposits.

The reemergence of distinct Placer Creek terraces near Mill Creek in lower Spanish Valley (Figs. 2 and 4) could reflect several factors. The shape and location of the deposits suggest that they may not be Pack Creek deposits, but rather are remnants of alluvial fans formed by Mill Creek because of its sudden change in gradient at the mouth of Mill Creek Canyon. Alternatively, the flow of Pack Creek may have been constricted by fans that extend northward from the southwest side of the valley, causing alluviation along the eastern edge of the valley (Fig. 2). These dissected, fan-shaped landforms are underlain by thin to coarse beds of interstratified cross-bedded sand, gravel, and unsorted angular deposits that probably represent either older side-stream fans or eolian deposits on bedrock. The deposits grade to the position of the Placer Creek surface as extrapolated by connecting terraces in the upper and lower valley (Fig. 4).

Another possible explanation for the vertically distinct terraces at both ends of Spanish Valley and the apparent downstream convergence of terraces in upper Spanish Valley is that Spanish Valley has experienced local subsidence. If the Placer Creek surfaces (Fig. 4), including the buried surface inferred in the middle of the valley, represent one valley profile, the middle portion of the valley has subsided since Placer Creek time. This subsidence could also have caused the progressive downvalley migration of fan deposition in upper Spanish Valley by effectively lowering the base level of the fans. Ongoing subsidence at the lower end of Spanish Valley may be indicated by

Figure 5. Buried soils in middle Spanish Valley. Location of site is shown in Figure 4. The arrow shows position of buried soil interpreted as being equivalent to soil developed on Placer Creek Formation.

marshes present along the Colorado River (Fig. 2). Subsidence in either area could be the result of tectonism, dissolution of salt below the caprock and surficial deposits in the valley, or migration of salt at depth. Colman (1983) documents major late Cenozoic geomorphic changes caused by movement of salt beneath Fisher Valley, located on the northern flank of the La Sal Mountains.

SOILS

Introduction

Calcic soils developed on Spanish Valley terraces provide a means of correlating and assessing the relative ages of the underlying Quaternary deposits. Progressive changes in the morphology and content of pedogenic calcium carbonate are particularly useful indicators of the ages of arid and semiarid soils (Machette, this

TABLE 2
MORPHOLOGIC STAGES OF CARBONATE ACCUMULATION IN SOILS

Stage	Gravelly Parent Material	Nongravelly Parent Material
I	Few, thin coats on clasts	Few filaments; flakes or coats on sand grains
II	Continuous coats on clasts; some interclast fillings; matrix somewhat whitened	Few to common nodules; nodules slightly to extremely hard; matrix slightly whitened
III	Continuous interpebble fillings; cemented and plugged horizon	Coalesced nodules; continuous cement in most of horizon
IV	Laminar horizon overlying plugged horizon	
V	Strong laminar horizon overlying plugged horizon; incipient brecciation and pisolith development	
VI	Strong brecciation and pisolith development	

Sources: Bachman and Machette, 1977; Shroba, 1977

volume; Bachman and Machette, 1977; Gile and Grossman, 1979). Changes in other soil parameters, including clay accumulation and reddening of the B horizon, were also investigated but were found to be less useful age indicators for Spanish Valley deposits. Soil horizon nomenclature generally follows that of the Soil Survey Staff (1975); terms pertaining to calcic soils are also discussed by Machette (this volume).

Methodology

Detailed profile descriptions, sampling of horizons, and laboratory analyses of particle-size distribution and carbonate content were made at seven backhoe excavations and one natural exposure of pre-Holocene terrace deposits (Fig. 2). Additional profile descriptions from natural exposures supplemented these detailed analyses. Standard Soil Conservation Service nomenclature (Soil Survey Staff, 1975) was used in designating soil horizons. The assignment of morphologic stages to calcic horizons (Table 2) follows the classification schemes of Bachman and Machette (1977) and Shroba (1977). Gravel percentages for horizons having more than 10 percent gravel (<2 mm) were determined for each horizon using a 50-point grid with a 5-cm spacing.

The particle-size distribution of the fine (≤2 mm) fraction of each horizon was measured by hydrometer and sieve analyses. Samples were pretreated to remove calcium carbonate ($CaCO_3$). Preliminary analyses of 10 samples showed that removal of organic matter, silica, and iron prior to particle-size analysis did not significantly affect the results, so these pretreatments were not performed during subsequent analyses. Moisture contents of the soils were measured during the analyses and were generally found to be less than 1 percent. The grain size classification follows the standard Soil Survey Staff (1975) classification. Analyses of samples with known size distribution and duplicate samples indicate that the laboratory data are generally reproducible within ±3 to 5 percent.

Calcium carbonate contents of samples were measured by two methods. A volumetric calcimeter (Chittick device) was used to measure the $CaCO_3$ content of fine-grained samples (Dreimanis, 1962). Acid neutralization (Black, 1965) was used to determine the $CaCO_3$ content of gravelly horizons and indurated samples. Results for duplicate analyses indicate that $CaCO_3$ measurements are generally reproducible to within ±5 percent.

To compare the degree of carbonate accumulation in profiles developed in different parent materials, weight-percent measurements were converted to mass of $CaCO_3$ per cm^2 of soil column (Birkeland, 1974). The mass of $CaCO_3$ per cm^2 of soil column for the whole profile is the cumulative total of the $CaCO_3$ content per cm^2 in each soil horizon. Channel samples for the $CaCO_3$ analyses were collected throughout the sampled profile with a maximum interval of 20 cm between samples.

Measurement or estimation of the bulk density of each horizon was necessary to make the conversions to mass. The bulk densities of fine-grained soil horizons and cemented petrocalcic horizons were determined by the soil aggregate method (Black, 1965) using air-dried natural peds or a small piece of calcrete. The soil aggregate method was also used to calculate the bulk densities of the matrix of gravelly horizons. The bulk density of coarse gravel clasts and their carbonate coatings was estimated to be 2.6 gm/cm^3. This estimate was based on several measurements of porphyritic clasts from different terraces, reported densities of sandstone and syenitic to dioritic rocks (Jackson, 1970), and the density of pure $CaCO_3$ (2.71 gm/cm^3). The value of 2.6 gm/cm^3 may be somewhat high for the bulk density of the coarse

TABLE 3. ACCUMULATION OF CALCIUM CARBONATE IN SOILS DEVELOPED ON QUATERNARY DEPOSITS IN SPANISH VALLEY AND WEST OF GREEN RIVER

Quaternary Formation or Deposit	Location[a]	No. of horizons analyzed from soil profile	Total CaCO3 in <2 mm fraction[c] (gm/cm²)	Estimated CaCO3 content of parent material[d] in <2 mm fraction (%;gm/cm²)	Total CaCO3 in >2 mm fraction (gm/cm²)	Estimated CaCO3 content of parent material[e] in >2 mm fraction (%;gm/cm²)	Total CaCO3 (gm/cm²)	Estimated total pedogenic CaCO3[f] (gm/cm²)
Younger Beaver Basin Fm	Sec 17, T27S R23E	5	9	5.5;8	24	10;16	33	9
Older Beaver Basin Fm	Sec 18, T27S R23E	8	7	2;3	7	3;4	14	7
Younger Placer Cr. Fm	Sec 16, T27S R23E	10	46	6.6;17	13	3;5	59	37
Older Placer Cr. Fm	Sec 20, T27S R23E	8	15	0.3;0.6	31	3;7	46	38
Older Placer Cr. Fm	Sec 18, T27S R23E	7	43	10;20	19	3.7;6	62	36
Middle Harpole Mesa Fm	Sec 22, T27S R23E	6	19	2.6;3	17	2.5;5	36	28
Middle Harpole Mesa Fm	Sec 28, T27S R23E	12	33	0.5;1	15	1;6	48	41
Older Harpole Mesa Fm	Sec 5, T27S R23E	8[d]	95	5;11	114	3.5;16	209	182
Gold Basin Fm (to depth of C-14 date)	Sec 22, T26S R22E	7[d]	4	1.5;3.4	-	-	4	0.6
800ft Green River terrace	Sec 11, T25S R16E	8[d]	63	10;16	66	10;14	129	99
Antelope Mesa[b]	Sec 8, T27S R16E			2;16				132
West of Green River		11[d]	144	8;51	-	-	144	93

[a]Location and soil descriptions of profiles are presented on Figures 3-17 and 3-24.

[b]Two different estimates used for CaCO3 content of parent material.

[c]The total mass of calcium carbonate expressed in grams of CaCO3 per square centimeter of soil area is computed by summing the CaCO3 mass in each horizon or sampling interval. Carbonate content (weight percentage, measured in laboratory analysis) is multiplied by bulk density (g/cm³) of each horizon and the horizon thickness (cm) to obtain the CaCO3 mass in each horizon (expressed as gm/cm² of soil column). Bulk densities of fine-grained material were measured by a water-volume-displacement technique using wax-coated soil clods. The original CaCO3 content of the parent material, estimated from the CaCO3 content at the base of the soil profile or nearby, equivalent parent material, was calculated for the soil column and subtracted from the total calculated CaCO3.

[d]Bulk densities estimated.

[e]Estimates based on laboratory analysis for the whole soil profile.

[f]CaCO3 content for horizons not analyzed were estimated from CaCO3 content of adjacent horizons.

fraction, because it does not take into account any pore space that may be present in the carbonate cement. However, values obtained for calcretes by Bachman and Machette (1979) are commonly in the range of 2.6 gm/cm³. The CaCO3 content of gravelly horizons was calculated by weighting the contents of the coarse and fine size fractions by the measured volumetric percent of each fraction in the horizon.

Spanish Valley Soil Chronosequence

Holocene Deposits. Gold Basin Deposits characteristically exhibit weak soil profile development. The alluvium commonly displays weak buried A horizons, distinguishable by their dark color, which appear to mark the top of individual flood deposits. Disseminated carbonate with weak Stage I forms are present at a few localities. An alluvial sequence that overlies a charcoal-bearing horizon dated at 1,280±55 years B.P. (Table 3) contains approximately 0.6 gm/cm² of soil column of pedogenic calcium carbonate. No distinct calcic horizon has formed at that site.

Pleistocene Deposits. Soils developed on pre-Holocene terraces in Spanish Valley typically exhibit complex profiles. At all sites examined, a relict soil with weak cambic and calcic horizons has formed on fine-grained deposits that overlie the gravelly alluvium. In deposits older than the Beaver Basin Formation, an older buried and subsequently truncated soil is represented by well-developed calcic or petrocalcic horizons in the upper portion of the gravel (Fig. 6; Fig. 7, profiles 3 through 8). Profiles developed on the Beaver Basin deposits (Fig. 7, profiles 1 and 2) may also be complex, but no clear evidence of discontinuities was observed.

Beaver Basin Formation. Two sampled soil profiles on deposits of Beaver Basin age show cambic or weak argillic horizons developed in fine-grained deposits and calcic horizons formed in the underlying gravelly alluvium (Figs. 6 and 7). The increased clay content in the upper portion of the profiles (Fig. 8, profiles A and B) may reflect parent material changes. Pedogenic reddening of the upper horizons was only seen in the older Beaver Basin deposits.

Figure 6. Fine-grained deposits of Gold Basin Formation overlying gravelly alluvium of the Beaver Basin Formation. Exposure is in the modern channel of Pack Creek. Surveying pole is approximately 2 m high.

The calcic horizons on the Beaver Basin Formation display continuous carbonate coats on clast bottoms, and nodules and filaments of carbonate are common in the matrix (Stage II morphology). The two sampled profiles contain 7 and 9 gm/cm^2 of pedogenic carbonate for the soil column. Comparison of the profiles supports Richmond's (1962) observation that the soils formed in the upper and lower members of the Beaver Basin Formation are not distinguishable.

We have interpreted the Beaver Basin profiles as a single-soil profile developed in both fine-grained and gravelly deposits. However, the calcic horizon in the gravel may represent a slightly older truncated soil, as appears to be the case in older soils.

Placer Creek Formation. Soils formed on the Placer Creek Formation at three sampling sites (Fig. 7, sites 3, 4, and 5; Fig. 9) are readily separable from those of the Beaver Basin Formation on the basis of carbonate accumulation. The calcic horizons of soils developed on Placer Creek deposits display continuous carbonate K fabric (Stage III), and at one locality (Fig. 7, site 5) laminar carbonate (Stage IV) overlies a cemented horizon containing more than 50 percent carbonate. The apparent variability in carbonate morphology on the two older Placer Creek deposits (sites 4 and 5) may be partly caused by more rapid carbonate accumulation, or by reduced leaching, at site 5 because

of its lower elevation. However, the pedogenic carbonate contents calculated for the three Placer Creek soil profiles are similar, ranging from 36 to 38 gm/cm^2 of soil column (Table 3).

No clear pattern of clay accumulation is apparent in soils formed on Placer Creek deposits (Fig. 9). A weak argillic (site 3) or cambic (site 5) horizon is preserved in the fine-grained deposits overlying the gravel, but the development of the B horizons seems anomalously weak compared to the degree of development of the underlying, well-developed calcic horizons. No horizon of increased clay content is present in the gravels at sites 3 and 4, but the high clay content in the K horizon at site 5 suggests that it may be part of an older B horizon engulfed by calcium carbonate (Gile and Grossman, 1979). We hypothesize that the general lack of argillic horizons associated with the calcic horizons in the gravel indicates that the B horizons were largely removed by erosion prior to deposition of the fine-grained material that overlies the gravel.

Harpole Mesa Formation. Calcic soils observed on the Middle Member of the Harpole Mesa Formation at Amasas Back (Fig. 7, sites 6 and 7; Figs. 10 and 11) display roughly the same thickness and stage of carbonate development as do soils on Placer Creek deposits at sites 3 and 4. Laminar carbonate, present in the soil developed in lower Placer Creek deposits at site 5, was

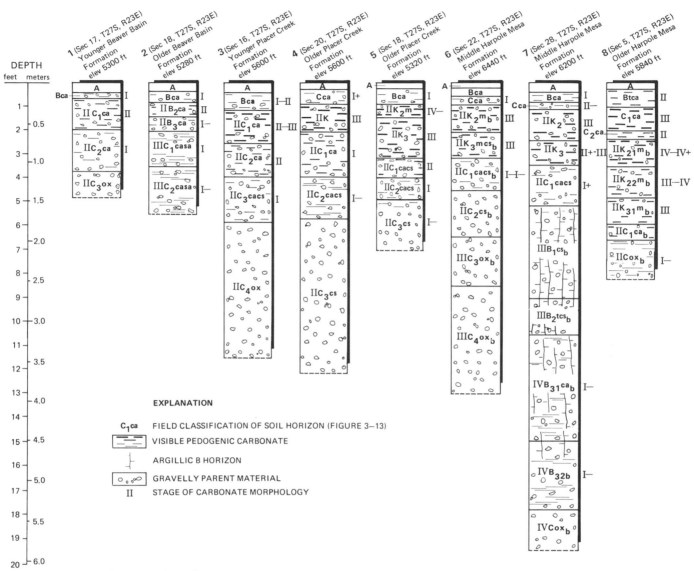

Figure 7. Soil profiles developed on terraces in Spanish Valley. Horizon nomenclature follows Soil Survey Staff (1975). Locations of profiles are shown in Figure 2.

not observed at sites 6 and 7. The pedogenic carbonate content at site 6 (28 gm/cm² for the soil column), is significantly lower than that of Placer Creek soils, and the carbonate content of profile 7 (41 gm/cm²) is only slightly greater (Table 3). As is the case with soils on the Placer Creek Formation, the B horizons associated with the calcic horizon in the gravel appear to have been eroded. Weak B horizons subsequently developed in the younger, fine-grained deposits overlying the gravel.

The anomalously weak development of soils on the gravel deposits on Amasas Back (profiles 6 and 7) can be interpreted in several ways. Microclimatic differences (due to its higher elevation) may have caused locally greater leaching. However, differences in elevation are only 200 to 400 m. The Amasas Back surface also appears to have been subject to complex, episodic deposition and (or) localized erosion. On the southwestern end of the surface, subtle topographic breaks resembling landslide

headscarps suggest that mass movement may also have disrupted the deposits. Soil development on the surface may have been periodically disrupted by deposition, or soil may have been removed by erosion since deposition of the middle Harpole Mesa Formation.

A buried clay-rich stratum that contains up to 29 percent clay underlies the carbonate soil at site 7 (Fig. 11). The clay content of this unit remains relatively high throughout the lower part of the profile, which suggests that the deposit could have been initially clay-rich. However, a 0.5-m thick zone with increased clay content may represent a buried argillic horizon. The profile at site 6 (Fig. 10) also shows high clay content in the lower horizons, but it lacks any zone of significant clay enrichment within those horizons. The presence of a buried soil beneath the calcic horizon would support the interpretation that Amasas Back is underlain by several deposits that span considerable time.

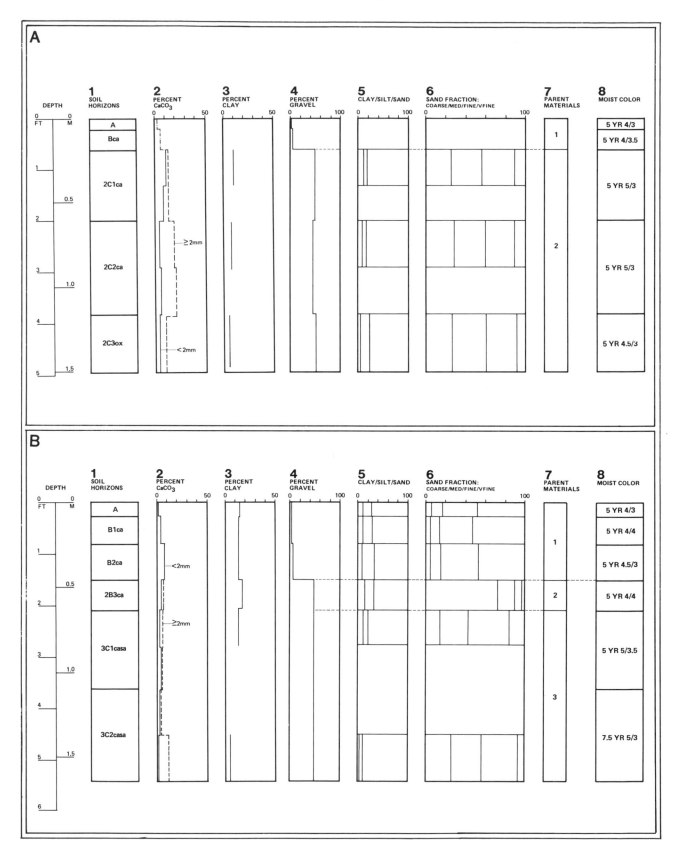

Figure 8. Laboratory and field data for Beaver Basin soil profiles. A: Younger Beaver Basin Formation (profile 1, Fig. 7); B: Older Beaver Basin Formation (profile 2, Fig. 7).

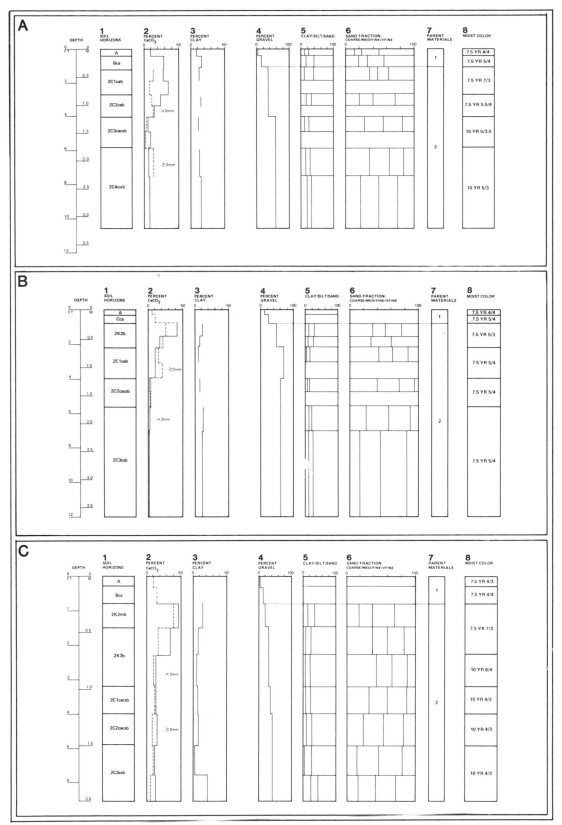

Figure 9. Laboratory and field data for Placer Creek soil profiles. A: Younger Placer Creek Formation (profile 3, Fig. 7); B: Older Placer Creek Formation (profile 5, Fig. 7); C: Older Placer Creek Formation (profile 4, Fig. 7).

Figure 10. Photograph showing soil developed on the middle Harpole Mesa Formation (profile 7, Fig. 7).

The soil observed on older Harpole Mesa alluvium at Johnson's-Up-On-Top (Fig. 7, site 8; Figs. 12 and 13) displays a far greater degree of carbonate accumulation than that seen on middle Harpole Mesa deposits. The presence of 1.5 m of indurated pedogenic K horizon displaying strong platy structure and laminar carbonate (Stage IV to IV+) suggests that the deposits may be as old as early Pleistocene (Machette, this volume). This age estimate is supported by paleomagnetic data from a sandy lens indicating that the deposit has reversed polarity. The profile contains about 182 gm/cm^2 of pedogenic carbonate in the soil column (Table 3). However, the content could have been increased at site 8, probably to a relatively minor degree, by case hardening on the natural scarp surface (Lattman, 1973).

A well-developed argillic horizon containing up to 25 percent clay is preserved in fine-grained eolian deposits overlying the gravel at Johnson's-Up-On-Top. This horizon has subsequently been engulfed by carbonate. The presence of well-developed argillic and calcic horizons on eolian deposits distinguishes this site from the other sampled profiles, where fine-grained deposits and soils appear to be of Holocene or late Pleistocene age.

DEPOSITS AND SOILS WEST OF GREEN RIVER

Early Pleistocene or older alluvial and eolian deposits occur on an extensive, irregular plateau south of Green River, Utah (Fig. 1). The deposits are 240 to 400 m above the present Green River channel and are capped by pedogenic calcrete (Fig. 14). In a typical exposure just northwest of Horseshoe Canyon, approximately 380 m above the present river channel, the soil has developed in 9 m of sand and pebbly sand that has reversed paleomagnetic polarity. A dense petrocalcic horizon displaying tabular structure, incipient brecciation, and an upper laminar zone (Stage V) forms the top 2.3 m of the deposit. Less continuous carbonate occurs to a depth of 4 m. This profile contains between 93 and 132 gm/cm^2 of soil column of pedogenic carbonate (Table 3).

Similarly well-developed calcic soils (Fig. 15) have been observed at other localities west of the Green River, including a gravel terrace only 240 m above the present Green River channel. At least 99 gm/cm^2 of carbonate have accumulated in this soil profile (Table 3). The apparently similar soil development on all

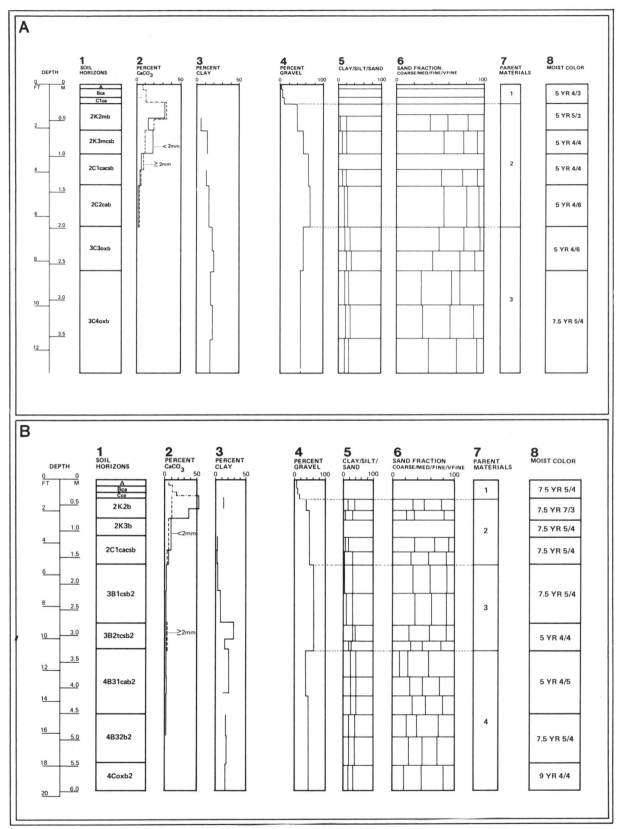

Figure 11. Laboratory and field data for soil profiles developed on middle Harpole Mesa Formation on Amasas Back. A: Middle Harpole Mesa Formation (profile 6, Fig. 7); B: Middle Harpole Mesa Formation (profile 7, Fig. 7).

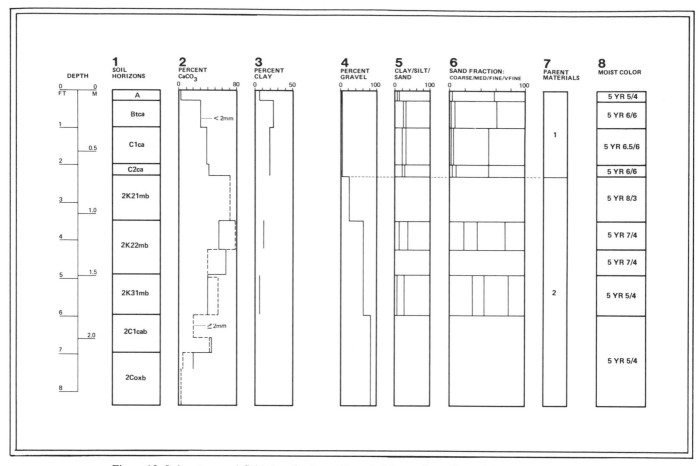

Figure 12. Laboratory and field data for lower Harpole Mesa soil profile at Johnson's-Up-On-Top. (profile 8, Fig. 7).

of these terrace deposits suggests that either the carbonate in the older deposits accumulated after the 240-m terrace had been cut, or that soils of different ages have attained a steady-state carbonate morphology so that they appear to be similar in age.

Pedogenic calcrete formed on the terraces west of the Green River is comparable in development to soils dated at least 0.5 m.y. old in Colorado, Nevada, and New Mexico (Bachman and Machette, 1977), and the soil may represent as much as 2 m.y. of development. Deposits on the 400-m terrace also display reversed magnetic polarity and are therefore at least 0.7 m.y. old.

AGE ESTIMATES OF DEPOSITS

As they contain few materials suitable for radiometric dating, Quaternary deposits in southeastern Utah are difficult to date. In addition to radiocarbon and paleomagnetic dating, we have investigated a number of techniques, summarized below, for obtaining relative or absolute ages of deposits.

Long-Term Rates of Carbonate Accumulation

If the average long-term influx rate of calcium carbonate is known for a region, ages of Quaternary deposits can be estimated from their pedogenic carbonate contents (Bachman and Machette, 1977; Machette, this volume). In using this method, it is assumed that (1) the long-term influx rate has been relatively constant during the Quaternary Period, (2) local variations in influx rates are relatively minor, (3) the climate and hence the extent of leaching have not fluctuated greatly, and (4) pedogenic carbonate has not been removed by erosion. Because these assumptions are never fully correct, the ages calculated by this method are only estimates.

Two deposits in the area that exhibit reversed magnetic polarity and well-developed soil profiles were used to estimate the maximum long-term carbonate accumulation rate for the region. Pedogenic carbonate contents of profiles west of the Green River (Fig. 15) and the profile at Johnson's-Up-On-Top in Spanish Valley (Fig. 7) were calculated. Assuming that these deposits are 0.7 m.y. old (the minimum age for magnetically reversed deposits), the maximum long-term accumulation rate is between 0.16 and 0.26 $gm/cm^2/1,000$ years (Fig. 16). Although the deposits that exhibit reversed polarity may be as old as 2.3 m.y., their geographic positions above present stream level suggest that the deposits are between 0.7 and 1.0 m.y. old. The lower

A

B

Figure 13. (A) View north toward Johnson's Up-On-Top. Spanish Valley is at left of photograph. Terrace is capped by alluvium of the lower Harpole Mesa Formation and is about 300 m above the Pack Creek channel. (B) K Horizon formed on the lower Harpole Mesa Formation at Johnson's-Up-On-Top (profile 8, Fig. 7).

of these maximum, long-term rates is comparable to that of the Beaver, Utah region (0.14 $gm/cm^2/1,000$ years) and to that calculated by Colman and others (in press) for Fisher Valley, Utah. The higher rate is similar to long-term accumulation rates for the Albuquerque (0.22 $gm/cm^2/1,000$ years) and Las Cruces (0.26 $gm/cm^2/1,000$ years) areas of New Mexico (Machette, this volume).

The high rate calculated for Johnson's-Up-On-Top can probably be considered an absolute maximum because the carbonate content of the sampled profile may partly reflect accumulation on the natural scarp surface (Fig. 13). In addition, the sampled locality exhibits the strongest calcrete development observed along the rim of the terrace surface.

Using the computed long-term accumulation rate for the Spanish Valley area, ages were estimated for younger Quaternary deposits in Spanish Valley (Fig. 16). The estimates obtained from the pedogenic carbonate contents of these deposits (Table 3) are in general agreement with ages based on regional correlation with glacial deposits in the Rocky Mountains (Richmond, 1962; Table 1). The low carbonate content of soils on Amasas Back (Fig. 7, profiles 6 and 7) may reflect erosion or microclimatic control of carbonate accumulated on that surface. The apparent anomalously high carbonate content of Holocene and late Pleistocene

Figure 14. Pedogenic calcrete west of the Green River (profile 3, Fig. 15).

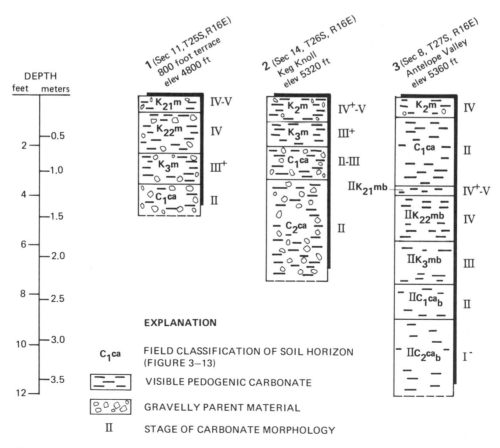

Figure 15. Soil profiles developed on alluvium and eolian deposits west of the Green River. Locations of profiles are shown in Figure 1.

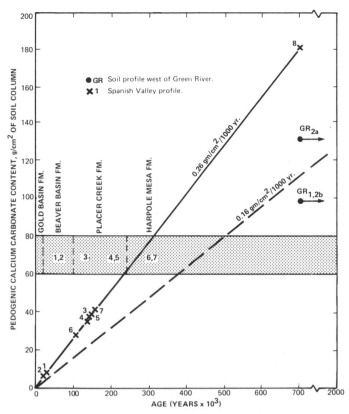

Figure 16. Long-term carbonate accumulation rates and age estimates for Spanish Valley deposits. The diagonal lines show maximum accumulation rates estimated from carbonate accumulation in three soil profiles developed on deposits having reversed polarity (GR 1 and 2b; GR 2a). Profiles from the Older Beaver Basin (X2), Younger Beaver Basin (X1), Middle Harpole Mesa (X6, X7), Older Placer Creek (X3, X4), Younger Placer Creek (X5), and Older Harpole Mesa (X8) Formations are plotted along the Spanish Valley accumulation trend line according to their pedogenic carbonate content (Table 3). The shaded band shows age estimates for the Spanish Valley formations based on regional correlation with glacial deposits in the Rocky Mountains (Table 1). Soil profiles are shown in Figures 7 and 15.

deposits probably reflects the short-term high influx rates of eolian $CaCO_3$ during this time. In spite of these variations in accumulation rates, the method appears to provide reasonable age estimates for Quaternary deposits in Spanish Valley.

Clast Weathering

Porphyritic igneous cobbles of intermediate composition were collected from subsurface soil horizons exposed in backhoe test pits excavated in the Beaver Basin, Placer Creek, and middle Harpole Mesa deposits. Measurements of oxidation rinds developed on these cobbles revealed no consistent trend with increasing age of the deposit. Variability in lithology, spalling of the outer surfaces of clasts, and deceleration of weathering caused by carbonate accumulation on the clasts probably are responsible for the lack of progressive oxidation-rind development. Additionally, the variable truncation of overlying soil horizons on the

terraces makes it difficult to compare the weathering histories of buried clasts. No grusified clasts were observed in deposits of the Beaver Basin Formation, whereas older deposits generally contain 10 to 30 percent grusified clasts in the soil profile. No consistent difference in grusification was detected between Placer Creek and Harpole Mesa deposits.

Amino-Acid Diagenesis

Amino-acid racemization has been a widely used tool for relative and, theoretically, absolute age determinations of fossil molluscs, foraminifera, and bone (Hare and others, 1980). More recently, investigators have suggested its use in detecting paleosols and possibly dating soils (Goh, 1972; Limmer and Wilson, 1980; Brigham, 1980). Because amino acids are of biological origin, they are present primarily in the L-stereoisomer configuration. Gradually the L-amino acids invert (or racemize) to the D-stereoisomer until a thermodynamically stable equilibrium mixture of D and L forms is reached. Thus, materials of increasing age should contain greater portions of D-amino acids. Specifically, the amount of interconversion (epimerization) of L-isoleucine into its diastereomer D-alloisoleucine is determined. Expressed as a ratio of D/L this fraction provides a measure of the extent of amino-acid diagenesis and a measure of relative age.

In soils, amino acids contained in organic material at the surface are absorbed into the crystalline structure of clay minerals where they become highly resistant to chemical or biological attack. Ideally, indigenous L-amino acids preserved in this manner slowly racemize during (and following) translocation of the clay particles down the soil profile and into the B horizon (Brigham, 1980). Because racemization reactions are strongly temperature-dependent, relative-age assignments are made with the least uncertainty if sites are located in the same microclimatic setting and the sediments are sampled from approximately 1 m or greater depth.

In order to assess the usefulness of the technique for relative soil-dating, amino-acid analyses were performed on soils from carbonate- or clay-rich horizons at the eight sampled profiles in Spanish Valley. The samples were collected from freshly exposed material at least 0.3 m below the ground surface. Analyses were performed by the Institute of Arctic and Alpine Research, University of Colorado, on the silt and smaller size-fraction, following dissolution of $CaCO_3$ and hydrolysis. Both the sediment residue and the supernatant liquid were analyzed; amino acids were most abundant in the sediment hydrolysate fraction.

The results of the analyses (Table 4) indicate that the organic material in near-surface K horizons is young. The low ratios may reflect contamination by newly infiltrated organic matter and the continuing accumulation of $CaCO_3$ in the calcic horizons. Ratios in the B horizon formed in the older Beaver Basin Formation (site 2) and the buried clay-rich stratum in the middle Harpole Mesa Formation (site 7) are significantly higher. These results suggest that clay may be more effective than carbonate in isolating organic matter. The high D/L ratio in the 3B1csb2

TABLE 4
AMINO-ACID RATIOS OF SPANISH VALLEY SOILS

Formation	Location	Horizon	Depth (cm)	D/L Ratio
Younger Beaver Basin	Sec 17, T27S, R23E-1	IIC1ca	40	.028
Older Beaver Basin	Sec 18, T27S, R23E-2	IIB3ca	60	.174, .180
Younger Placer Creek	Sec 16, T27S, R23E-1	IIC1ca	50	.057
Older Placer Creek	Sec 20, T27S, R23E-1	IIK2	50	.051
Older Placer Creek	Sec 18, T27S, R23E-1	IIK2m	30	.031
Middle Harpole Mesa	Sec 22, T27S, R23E-1	IIK2mb	30	.035
Middle Harpole Mesa	Sec 28, T27S, R23E-1	IIK2b	50	.057
Middle Harpole Mesa	Sec 28, T27S, R23E-1	IIIB1csb2	250	.45
Older Harpole Mesa	Sec 5, T27S, R23E-2	IIK22mb	110	.03

horizon of the middle Harpole Mesa deposits may also support the interpretation that the horizon is a paleosol developed on an older buried deposit.

Uranium-Thorium Dating

Researchers in southeastern California (Ku and others, 1979; Bischoff and others, 1981) have successfully used uranium-thorium dating to obtain ages for pedogenic carbonate rinds formed on gravel clasts. Prior to this investigation, a single Th^{230}-U^{234} date was obtained from the carbonate rind on a cobble from the Placer Creek deposit at the mouth of Mill Creek Canyon (Fig. 2). The analysis, performed by T. L. Ku of the University of Southern California, produced a probable date of 15,000±1,000 years B.P.

The uranium-thorium date appears to be significantly younger than the age suggested by the degree of soil development observed on the terrace. The calcic horizon formed on the deposit displays Stage II to III morphology, and is about 0.8 m thick (Fig. 17). An argillic B horizon (sandy clay loam) that is as much as 0.5 m thick exhibits strong blocky structure and has developed in fine-grained deposits overlying the calcic horizon. The degree of soil development is therefore comparable to, or only slightly less than, that observed at the sampled Placer Creek sites (Fig. 7). This supports Richmond's (1962) assignment of these deposits to the Placer Creek Formation.

The probable uranium-thorium dates obtained from two other samples, collected from terraces in the vicinity of Blanding and Bluff, Utah, south of Moab, were also anomalously younger than the ages suggested by either the elevation of the terraces above present stream levels or, for one of the two samples, by the degree of calcic soil development on the deposits. We have inter-

preted these consistently young ages as indicating that the carbonate rinds are contaminated with younger $CaCO_3$ that was incorporated after the main part of the soil had formed. In order to determine whether this problem is an artifact of sampling methods or sites, or is typical of pedogenic carbonate in the area, multiple analyses from several sites need to be made.

CORRELATION OF DEPOSITS

Our correlation of the Quaternary deposits of Spanish Valley (Table 1) generally supports Richmond's (1962) relative chronology developed for the La Sal Mountains. On the basis of calcic soil development, the Beaver Basin Formation appears equivalent to Pinedale-aged deposits of the Rocky Mountain region (Meierding and Birkeland, 1980). Carbonate buildup in soils on Placer Creek deposits is equivalent to that reported for many Bull Lake deposits (Shroba, 1977). On the basis of pedogenic carbonate contents and constant long-term carbonate influx rates, the soils formed on Beaver Basin deposits are between 27,000 and 64,000 years old (Fig. 16). Those formed on Placer Creek deposits are between 138,000 and 271,000 years old (Figure 16). The significant difference in pedogenic carbonate content between Beaver Basin and Placer Creek deposits (Table 3; Fig. 16) supports Pierce and others' (1976) interpretation that Bull Lake deposits are significantly older than Pinedale deposits.

The soil developed on the middle Harpole Mesa Formation at Amasas Back is not significantly more developed than that on Placer Creek deposits. However, based on relative soil development, the lower Harpole Mesa Formation is clearly much older than the middle Harpole Mesa Formation or younger deposits. Differences in microclimate at the Placer Creek and middle Harpole Mesa sites may have resulted in different rates of carbonate

Figure 17. Placer Creek terrace at the mouth of Mill Creek Canyon.

accumulation, and Amasas Back may represent a depositional or truncated erosion surface that developed on deposits of both Placer Creek and middle Harpole Mesa age.

In the middle part of Spanish Valley, where gravel pits are numerous, a complex pattern of buried soils and deposits was observed beneath an almost planar topographic surface. Additional information derived from subsurface measurements regarding alluvial deposits and buried soils in the middle portion of Spanish Valley could aid in correlating terraces and understanding the Quaternary history of the valley. Data were insufficient to compile a profile of the basal alluvial contact or of disconformities within the alluvium from existing water-well logs. Such a profile would aid in identifying possible subsidence in segments of the valley.

ACKNOWLEDGMENTS

The material presented in this paper is a portion of the results to date of ongoing geologic studies in the Paradox basin in Utah. The investigations are being conducted by Woodward-Clyde Consultants (WCC) for the Office of Nuclear Waste Isolation (ONWI), Battelle Memorial Institute, Columbus, Ohio, prime contractor to the Department of Energy. The project is under the direction of Fred Conwell, WCC's project manager. In addition to the authors, Kathryn Hanson and Roy McKinney of WCC also participated in field investigations, and F. H. Swann, III reviewed the study sites. Peter W. Birkeland of the University of Colorado also reviewed field interpretations of soils in Spanish Valley. Jerry Kuns assisted in the development of techniques for measuring bulk densities.

Laboratory analyses of soils were performed by Nelson Laboratories of Stockton, California. We are especially grateful to Jack Shimasaki, Matt Buchwitz, Roger Buchwitz, and Mike Purser for their careful work and willingness to experiment with analytical techniques. Radiocarbon analyses were performed by Irene Stehli of the Dicarb Radioisotope Company, Gainesville, Florida. Paleomagnetic measurements were made by Duane Packer and Jeff Johnston at WCC's paleomagnetism laboratory. T-L. Ku of the University of Southern California performed uranium-thorium analyses. Sample preparation, amino-acid analyses, and interpretation of results were made by Julie K. Brigham, Gifford Miller, and William McCoy at the Amino Acid Geochronology Laboratory at the University of Colorado's Institute for Arctic and Alpine Research, Boulder, Colorado. Michael N. Machette of the U.S. Geological Survey supplied two $CaCO_3$ analyses of modern dune sand.

We also appreciate the critical review of this manuscript by Terry Grant (WCC), N. A. Frazier (ONWI), and Steve Colman (U.S. Geological Survey). The help of Carol Droge in typing the manuscript, V-Anne Chernock, project editor, and WCC's drafting staff is also gratefully acknowledged.

Several colleagues, including Jennifer W. Harden and Michael N. Machette of the U.S. Geological Survey, have supplied suggestions and advice during the course of this project. Finally, we would like to acknowledge the encouragement and advice of the late Denis Marchand who freely shared his ideas and suggestions.

REFERENCES CITED

Bachman, G. O., and Machette, M. N., 1977, Calcic soils and calcretes in the southwestern United States: U.S. Geological Survey Open-File Report 77-794, 163 p.

Birkeland, P. W., 1974, Pedology, weathering, and geomorphological research: New York, Oxford University Press, 285 p.

Bischoff, T. L., Shlemon, R. J., Ku, T. L., Simpson, R. D., Rosenbaurer, R. J., and Budinger, F. E., Jr., 1981, Uranium-series and soil-geomorphic dating of the Calico archaeological site, California: Geology, v. 9, p. 576–582.

Black, C. A., ed., 1965, Methods of soil analysis, part 1: Physical and mineralogical properties, including statistics of measurement and sampling: American Society of Agronomy, Inc., Agronomy Series No. 9, 770 p.

Brigham, J. K., 1980, The epimerization of isoleucine and the relative concentration of amino acids in paleosols: a review and pilot study: unpublished report, University of Colorado, 49 p.

Colman, S. M., 1983, Influence of the Onion Creek salt diapir on the late Cenozoic history of Fisher Valley, southeastern Utah: Geology, v. 11, p. 240–243.

Colman, S. M., Choquette, A. F., and Hawkins, F. F., in press, Physical, soil, and paleomagnetic stratigraphy of the upper Cenozoic deposits in Fisher Valley, Utah: U.S. Geological Survey Bulletin.

Dreimanis, A., 1962, Quantitative determination of calcite and dolomite by using Chittick apparatus: Journal of Sedimentary Petrology, v. 32, p. 520–529.

Gile, L. H., and Grossman, R. B., 1979, The desert soil project monograph: U.S. Department of Agriculture, Soil Conservation Service, 984 p.

Goh, K. M., 1972, Amino acid levels as indicators of paleosols in New Zealand soil profiles: Geoderma, v. 7, p. 33–47.

Hare, P. E., Hoering, T. C., and King, K., Jr., 1980, Biogeochemistry of amino acids: New York, John Wiley and Sons, 558 p.

Hunt, C. B., 1956, Cenozoic geology of the Colorado Plateau: U.S. Geological Survey Professional Paper 279, 99 p.

Iorns, W. V., Membree, C. H., and Oakland, G. L., 1965, Water resources of the upper Colorado River Basin—Technical Report: U.S. Geological Survey Professional Paper 441, 370 p.

Jackson, K. C., 1970, Textbook of lithology: New York, McGraw-Hill Book Company, 552 p.

Ku, T. L., Bull, W. B., Freeman, S. T., and Knauss, K. G., 1979, Th^{230}-U^{234} dating of pedogenic carbonates in gravelly desert soils of Vidal Valley, southeastern California: Geological Society of America Bulletin, Part 1, v. 90, p. 1063–1073.

Lattman, L. H., 1973, Calcium carbonate cementation of alluvial fans in southern Nevada: Geological Society of America Bulletin, v. 84, p. 3013–3028.

Limmer, A. W., and Wilson, A. T., 1980, Amino acids in buried paleosols: Journal of Soil Science, v. 31, p. 147–153.

Machette, M. N., 1985, Calcic soils of the American Southwest, in Weide, D. L., ed., Soils and Quaternary geology of the southwestern United States: Geological Society of America Special Paper 203 (this volume).

Madole, R. F., and Shroba, R. R., 1979, Till sequence and soil development in the North St. Vrain drainage basin, east slope, Front Range, Colorado: in Ethridge, F. G., ed., Field guide, Northern front range and northwest Denver basin, Colorado, Fort Collins, Colorado State University, p. 124–178.

Meierding, T. C., and Birkeland, P. N., 1980, Quaternary glaciation of Colorado, in Kent, H. C., and Porter, K. W., eds., Colorado Geology: Rocky Mountain Association of Geologists 1980 Symposium, p. 165–173.

Pierce, K. L., Obradovich, J. D., and Friedman, I., 1976, Obsidian hydration dating and correlation of Bull Lake and Pinedale glaciations near West Yellowstone, Montana: Geological Society of America Bulletin, v. 87, p. 703–710.

Richmond, G. M., 1962, Quaternary stratigraphy of the La Sal Mountains, Utah: U.S. Geological Survey Professional Paper 324, 135 p.

Shackleton, N. J., and Opdyke, N. D., 1973, Oxygen isotope and paleomagnetic stratigraphy of equatorial Pacific core V28-238: Oxygen isotope temperatures and ice volumes on a 10^5 year and 10^6 year scale: Quaternary Research, v. 3, p. 39–55.

Shroba, R. R., 1977, Soil development in Quaternary tills, rock-glacier deposits and taluses, southern and central Rocky Mountains [Ph.D. thesis]: Boulder, University of Colorado, 424 p.

Shroder, J. F., Jr., Giardino, J. R., and Sewell, R. E., 1980, Tree-ring and multi-criteria, relative-age dating of mass movement and glacial phenomena, La Sal Mountains, Utah: American Quaternary Association Abstracts and Programs, Sixth Biennial Meeting, p. 175.

Soil Survey Staff, 1975, Soil taxonomy: A basic system of soil classification for making and interpreting soil surveys: U.S. Department of Agriculture, Soil Conservation Service, Agricultural Handbook No. 436, Washington, D.C., U.S. Government Printing Office, 754 p.

Sumsion, C. T., 1971, Geology and water resources of the Spanish Valley area, Grand and San Juan Counties, Utah: Utah State Department of Natural Resources Technical Publication 32, 45 p.

Williams, P. L., 1964, Geology, structure, and uranium deposits of the Moab quadrangle, Colorado and Utah: U.S. Geological Survey Miscellaneous Geologic Investigations Map I-360, scale 1:250,000.

Manuscript Accepted by the Society January 12, 1985

Geological Society of America
Special Paper 203
1985

Smectitic pedogenesis and late Holocene tectonism along the Raymond Fault, San Marino, California

Glenn Borchardt
California Division of Mines and Geology
Ferry Building
San Francisco, California 94111

Robert L. Hill*
California Division of Mines and Geology
Los Angeles, California 90012

ABSTRACT

An organic rich, gleyed soil (Cumulic Haplaquoll) on the edge of a sag pond near San Marino High School illustrates the effects of poor drainage and tectonism on soil formation (pedogenesis) in a semiarid region during the Holocene. Combined with other characteristics of soil development, the organic matter masks the presence of six depositional strata in the granodioritic alluvium in which the soil formed. Two of these units occur only on the south side of a strand of the Raymond fault, three occur only on the north side, and one occurs on both sides.

The dominant clay minerals are smectites that have formed from soil solution within the past 10,000 years. Beidellite formed in horizons with exchangeable aluminum (pH less than 6.5), and montmorillonite formed in horizons without exchangeable aluminum (pH greater than 6.5). A soil tongue produced by fault movement was relatively unweathered after it was emplaced by an earthquake that occurred less than 1,400 years ago. Compared to other Holocene soils, pedogenesis at this site is intermediate between the weakly developed Hanford soils formed in granitic alluvium in the San Joaquin Valley and the strongly developed Concepcion soils formed in continental terrace deposits along the coast.

INTRODUCTION

In semiarid regions, depressions produced by fault movement provide unique environments for pedogenesis under restricted drainage. We had an opportunity to study an organic-rich, gleyed soil (Cumulic Haplaquoll) adjacent to a sag pond as part of an investigation of the age and activity of the Raymond fault (Figs. 1 and 2). The site was excavated by LeRoy Crandall and Associates, who concluded that the fault was active during the Holocene and that plans for a proposed sports facility would have to take this into account (Bryant, 1978). Discovered beneath the asphalt near San Marino High School (Fig. 3), this soil

had a very dark, thick A horizon that obscured subtle differences among the numerous alluvial units in which we supposed it had formed. The lower soil horizons were obviously offset by a strand of the fault (Fig. 4), but the techniques of soil science were necessary for delineating hidden strata in the soil horizons in order to determine the age of the most recent tectonic movement.

Recent work has demonstrated the power of physical, geochemical, and mineralogical methods for dating and interpreting fault features in soils (Swan and Hanson, 1977; Borchardt, Taylor and Rice, 1980; Borchardt, Rice, and Taylor, 1980; Borchardt and others, 1982). In applying these techniques we first evaluated soil-survey and geological reports concerning the characteristics and ages of the soils and sediments of the area.

---

*Present address: KOCH Exploration Co., 2716 Ocean Park Blvd., Santa Monica, California 90405.

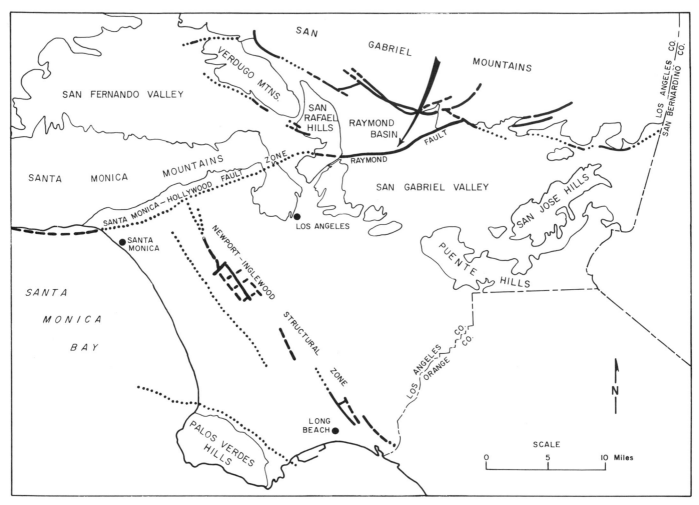

Figure 1. Los Angeles area showing the Raymond fault between the Raymond Basin and San Gabriel Valley and approximate location (arrow) of the San Marino High School trench (Hill and others, 1978).

The major points we found applicable to this study were: (1) most of the soils in the area are well-drained and are dominated by kaolinitic clays, and (2) the parent material at the site is Holocene alluvium derived from granodiorite of the San Gabriel Mountains. After logging and describing the fault, we sampled two profiles that were about 1 m on either side of a "soil tongue" or infilling that we suspected to be of tectonic origin (Fig. 4). Material was gathered for carbon-dating the soil tongue and the overlying strata that appeared to be undisturbed. Next, we tried to establish the nature of the initial material from which the soil formed. In this part of our study particle-size distribution data yielded coarse-sand/fine-sand and silt/sand ratios that helped delineate distinct alluvial strata obscured by soil development.

Our main concern is pedogenesis. We note the pattern of organic matter and clay accumulation as a function of the textural and mineralogical composition of the initial materials. The discovery that smectites (expanding 2:1 layer silicate clays) have formed from soil solutions at this site during the Holocene is of great significance for dating soils in similar environments. Also of

special interest to clay mineralogists is the observation that neogenetic beidellite (a tetrahedrally substituted smectite) appears to form in acidic horizons, while neogenetic montmorillonite (an octahedrally substituted smectite) appears to form in neutral horizons. The knowledge that these clay minerals have formed in situ helps establish that, except for tectonism, the deposits at the site have been physically undisturbed since their deposition.

Finally, we applied the above information in evaluating the recency and style of faulting. Compared to other samples, the soil in the tongue found between the two profiles had undergone little pedogenesis since its emplacement. Its interpretation as a relatively young feature was supported by soil carbon dates indicating that the event that produced the tongue probably occurred less than 1,400 years ago.

SOILS OF SAN MARINO

The soil in the area of the San Marino trench is mapped as Hanford fine sandy loam (Eckmann and Zinn, 1917). It is a circular polypedon surrounded by Ramona loam on the south and Ramona gravelly loam on the north. Eckmann and Zinn,

Figure 2. Oblique aerial photograph taken in 1924 of the area east of Rubio Wash and west of San Gabriel Boulevard; view northwest. In 1924, subtle fault-related features could be observed. A subtle scarp and evidence of the contrast in soil moisture can be seen just east of Rubio Wash (modified from Bryant, 1978). Photo courtesy of Geography Department, University of California, Los Angeles, Spence Air Photo Collection).

p. 41, describe the Hanford fine sandy loam in the area as "brown or light grayish fine sandy loam, 20 to 30 inches in depth, underlain to a depth of more than 6 feet by a subsoil which differs little from the surface material." Like the Hanford soils developed from granitic alluvium in the San Joaquin Valley (Harden and Marchand, 1977, p. 36), these soils lack an argillic B horizon.

Hanford soils are found on large alluvial fans at the mouths of major canyons leading into the Raymond Basin from the east. Soil on the older alluvial fans in the west side of the basin is mapped chiefly as the Ramona series, which has coarser parent materials and redder colors. Ramona loam, for example, has a subsoil which is "compact loam or clay loam . . . showing some weathering and stratification. An iron-clay hardpan is exposed in places . . ." (Eckmann and Zinn, 1917, p. 33). Although the soils

in the trench here are undoubtedly Holocene, they are loam in texture—not fine sandy loam like the Hanford series. As a side issue, we hoped to determine if the extra heavy texture is the result of clay that was reworked from the older Ramona series or if the clay had formed in situ.

ALLUVIAL DEPOSITION

Bedrock Sources

The alluvial sediments in the San Marino area were derived for the most part from intrusive igneous rocks and to a lesser extent from metamorphic rocks that constitute the San Rafael Hills to the west and the San Gabriel Mountains to the north and

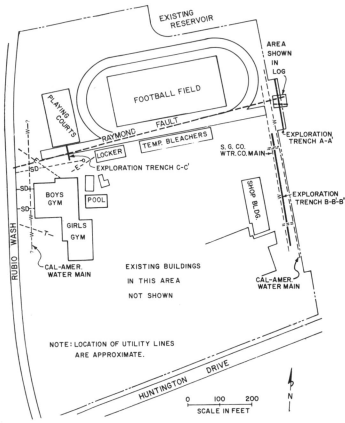

Figure 3. San Marino High School plot plan showing the location of exploration trenches (from Bryant, 1978). Map courtesy of LeRoy Crandall and Associates, Los Angeles.

east (Fig. 1). The igneous rocks are generally dioritic and include the Mt. Lowe Granodiorite which supplies the most resistant clasts found in sediments of the area. The crystalline rocks from which the sediments were derived are rich in feldspathic minerals that readily decompose to form clay minerals. From water-well logs, Buwalda (1940, p. 16) estimated that about half, or more, of the volume of the sedimentary fill within the Raymond Basin is clay derived from kaolinization of feldspars originally deposited as sand and gravel. These clayey sediments comprise the older alluvium that is presently exposed in uplifted fault blocks along the Raymond fault and in canyons along streams that have become entrenched into the deposits of the basin.

Alluvial Sequence

Except for the near-surface deposits, such as those exposed in the study area, the sediments beneath San Marino are part of the older alluvium considered to be pre-Holocene. The older alluvium is up to 300 m thick and generally is capped by soils of the Ramona series. The younger sediments were transported to the area by Eaton Wash and its distributaries to the east of the trench site (Fig. 2). These younger sediments, which have Hanford soils developed in them, constitute a relatively large alluvial fan having gentle south-sloping gradients. Eaton Wash is pres-

ently in a period of down-cutting into the deposits of the Raymond Basin and no longer supplies sediment to the San Marino area.

Disruption of drainage along the Raymond fault has resulted in local reworking of older deposits. This is evident from present topography along the fault zone and from the nature of the sediments exposed in trenches across the trace of the fault. Well-sorted quartz sands commonly occur immediately south of the main trace of the fault. It is possible that some of these are reworked from the older alluvial deposits derived from the upthrown block north of the fault trace. Some of these sands have been tectonically disturbed as a result of liquefaction and extrusion as sand boils. Sag-pond deposits consisting of interbedded or thin-bedded sand, silt, clay, and peat, or deposits of organic silt and clay several feet in thickness have been exposed in trenches at San Marino High School (C-C', just south of the playing courts in Fig. 3), at Sunny Slope Reservoir to the east (Hill and others, 1978; Crook and others, 1978), and in cores from Lacy Park to the west.

Sag-Pond Evolution

Our study was conducted at the south edge of what appears to have been a sag pond formed in a graben between two strands of the Raymond fault. The pond no longer exists, but we suspect that it once had the general outlines of the football field northwest of our sample site (Fig. 3). Historical evidence for the sag pond was mentioned by Buwalda (1940, p. 53), who placed it just east of Rubio Wash, south of the area of color contrast, and north of the strand we evaluated (Figs. 2, 3). Deposits characteristic of sag ponds or marshy areas were encountered north of the fault trace exposed in trench C-C' (Fig. 3). A detailed log of this trench is in Hill and others (1977); a generalized log is in Bryant (1978). Buwalda (1940, p. 51–54 and p. 124–128) also mentioned evidence for a second fault trace located parallel to and as much as 100 m north of the trace we studied. From the nature of the deposits exposed in trench C-C' and from the down-warping of the beds south of the shear zones exposed there, it appears that the block between the two faults has been down-dropped relative to the blocks to the north and south. Although left-lateral reverse movement on the Raymond fault produces regional uplift on the north, the trace we studied underwent some normal movement, down to the north (Hill and others, 1977, 1978; Crook and others, 1978).

METHODS

The trench studied in this paper, labeled A-A' in Fig. 3, was dug and shored by LeRoy Crandall and Associates on August 22, 1977. The east wall of the trench was logged by Robert L. Hill and David R. Smith (Fig. 4) using a level rectangular grid with 30 cm by 30 cm spacing. Soil profiles 2 m south (profile A) and 1.5 m north (profile D) of the northern-most fault break in the trench were described and sampled by Glenn Borchardt (Fig. 4

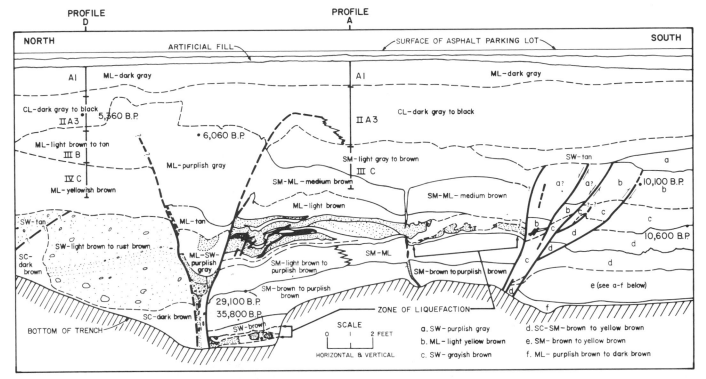

Figure 4. Log of the east side of trench A-A′ at San Marino High School (modified from Bryant, 1978); the detailed log is in Hill and others (1977), plate 3. The two-letter particle size-expansivity designations are based on the Unified Soil Classification System (Borchardt, 1977c, p. 257 and Soil Survey Staff, 1975). Profiles A and D and the carbon dates (from Crook and others, 1978, p. 76) are projected from the west side of the trench. An exception was the 10,100-year-old carbonaceous silt we obtained from the east wall (analysis by C. R. Berger, U.C.L.A. Radiocarbon Laboratory, Los Angeles).

and Table 1). Complete vertical sections were sampled at approximately 5-cm intervals, and samples centrally located within important soil horizons were selected for laboratory analysis. A soil tongue lying directly over the main shear zone was sampled at 62 to 67 cm, 92 to 97 cm, and 102 to 127 cm. The samples were analyzed in the laboratory using the methods described below:

Particle-Size Distribution

Particle-size distribution was determined by the hydrometer method (Day, 1965), but using sodium metasilicate instead of calgon (Borchardt, 1977a, p. 9). Carbonates were removed by treatment with pH 5 sodium acetate. Disperson was achieved with mild hydrogen peroxide treatment and an Iowa jet dispersion apparatus (Chu and Davidson, 1953).

X-ray Diffraction of the Clay Fractions

Clay fractions (<2 μm) obtained in the above procedure were saturated with magnesium and used to prepare slides by the smear-on-paste technique (Theisen and Harward, 1962). A 15 cm flexible plastic rule was used instead of a spatula for spreading the paste (Borchardt, 1977a, p. 9). Slides were air dried for one hour, placed in a 54 percent humidity chamber overnight, and

x-rayed in a controlled 54 percent relative humidity environment. Under these conditions, mica would have been detected by the presence of 10Å peaks. Next, the sample was heated at 110°C for two hours in glycerol vapor, which was then allowed to condense onto the sample overnight (Brown and Farrow, 1956). Another Mg-saturated sample was heated at 60°C for two hours in ethylene glycol vapor that was also allowed to condense overnight. X-ray patterns for ethylene glycol treated slides of montmorillonite and beidellite standards were identical, showing expansion to about 17Å. Beidellite was identified by its resistance to expansion when solvated with glycerol vapor (Harward and Brindley, 1964).

K-saturated slides were similarly prepared and heated to 110°C and 550°C. Vermiculite would have been indicated by collapse of 14Å peaks to 10Å after K saturation, drying at 110°C, and x-raying at 0 percent humidity. Traces of kaolin (kaolinite or halloysite) were indicated by the disappearance of a 7Å peak after heating the K-saturated sample for three hours at 550°C. Chlorite would have been indicated by the presence of a 14Å peak after 550°C heating.

Determination of pH

The pH was determined by mixing the soil with water until

it had the consistency of a paste. A Chemtrix Type 60A pH meter was used for the determination.

Total Iron and Calcium

Fe and Ca were determined by using a Phillips x-ray spectrometer and a method of sample preparation modified from Borchardt and Theisen (1971). Briefly, air-dried soil was sieved through a No. 40 sieve (420 μm) and then added to a 32 mm aluminum Spex cap containing polyvinyl alcohol as a backing material. This was pressed in a die assembly at 2300 kg/cm^2 to form a smooth surface.

Total Fe was calculated from the count rate of the Fe^{++} peak at 51.75°2θ. The chromium x-ray tube was operated at 45 kv and 25 ma. X-rays were detected with a LiF crystal and scintilation counter operated in air at 960 volts. A pulse height analyzer was used with the baseline at 4.0 volts, and the window at 20.0 volts.

Total Ca was calculated from the count rate of the Ca^{++} peak at 113.10°2θ. The chromium x-ray tube was operated at 30 kv and 10 ma. X-rays were detected with a LiF crystal and flow-proportional counter operated in a less than 300 μm vacuum at 1,530 volts. A pulse height analyzer was used with the baseline at 4.0 volts, and the window at 11.0 volts.

The count rates for Fe and Ca were compared directly to those for a standard andesite (USGS-AGV-1) that contains 4.73 percent Fe and 3.50 percent Ca (Flanagan, 1973, p. 1190).

Color

Soil colors were determined by comparison to the standard Munsell soil color charts. When a color name is given, the name refers to the first cited notation in parentheses (Soil Survey Staff, 1951, p. 203). For example, brown (10YR5/3d, 2/2m) refers to the dry color, brown (10YR5/3d), and the moist color, very dark brown (10YR2/2m). This notation for dry (d) and moist (m) colors is becoming more common in the literature (e.g., Buntley and others, 1977, p. 401) and avoids confusion in preparing field descriptions.

PARENT-MATERIAL HETEROGENEITY

The generalized log of trench A-A' depicts many thin alluvial strata at depths between 2 and 3 m. (Fig. 4). Obviously undisturbed by soil formation, these units indicate the texturally heterogeneous nature of the alluvium in which the soils above must have developed. The strata are sheared in numerous places and some of them have undergone liquefaction; this suggests that the slip on the Raymond fault is seismic rather than aseismic.

Throughout the section, the alluvium contains rounded, saprolitic clasts derived from the Mt. Lowe Granodiorite. In the lower strata, even cobbles greater than 15 cm in diameter can be crushed by hand. These saprolitic clasts diminish in size upward toward the soil surface; in the upper 10 cm of the soil they are

entirely absent. The soil itself contains almost no gravel-size particles strong enough to survive the rigors of particle-size analysis (Tables 1 and 2). The ubiquity of weathered granodioritic clasts suggests that the provenance was the same for all of the sedimentary strata exposed in the trench.

Coarse-Sand/Fine-Sand Ratios

Laboratory analysis of particle-size distribution was used to assess parent-material heterogeneity. Because of its small surface area, the sand fraction is generally the least weathered of the fine (<2 mm) materials in soils (Borchardt and others, 1968). The ratio of the sand-size fraction in one soil to the sand-size fraction in another thus remains relatively constant for long periods of time. Variation, if any, may result from the original particle-size distribution of the alluvial units from which the soil formed.

In trench A-A', the ratio of the percentage of coarse sand (2.00–0.25 mm) to the percentage of fine sand (0.25–0.05 mm) does not appear to reflect soil weathering (Fig. 5). In profile A the coarse-sand/fine-sand ratio generally increases with depth, while in profile D it decreases with depth. Thus the depositional heterogeneity that was obvious in the lower, unweathered strata is also found within the soil horizons where it is visually masked (dashed lines in Fig. 4) by soil development. Only the A1 horizon seems to be similar on both sides of the fault.

Silt/Sand Ratios

Silt/sand ratios convey two types of information about a soil. Changes in the silt/sand ratio may indicate heterogeneities in the original parent material, as is the case with the coarse-sand/fine-sand ratios. Alternatively, soil weathering may result in an increase in silt through physical or chemical breakdown of the sand particles. In general, changes in the silt/sand ratio due to soil formation should vary gradually with depth, whereas changes due to stratification should be abrupt. Unlike the clay fraction, neither silt nor sand is normally translocated to lower horizons in medium-textured soils (Borchardt and others, 1968), therefore an increase in silt exceeding that of the parent material should indicate in situ weathering.

In the IIA3 horizons of profiles A and D, the fact that silt/sand ratios increase slightly with nearness to the surface could be a result of weathering (Fig. 6). However, in the IIIB horizon of profile D, silt/sand ratios actually decrease with nearness to the surface. Similarly, the sand contents (Table 2) in this zone are less than they are elsewhere in the profile; this is another indication that the IIIB soil horizon may have been deposited as a distinct alluvial unit. The physical or chemical breakdown of sand particles into silt particles is not a significant process in this soil.

In sum, laboratory data suggest that profiles A and D formed within fine-grained alluvium derived mostly from the Mt. Lowe Granodiorite. The depositional processes, clearly discerned from beds of varying texture in the unweathered parent material, persisted long enough to produce the upper 1 to 2 m of the

TABLE 1. ABBREVIATED SOIL PROFILE DESCRIPTIONS FOR THE WEST WALL OF
TRENCH A-A' AT SAN MARINO HIGH SCHOOL*

Horizon	Depth (cm)	Description
		Soil Profile A (2 m south of the main fault break)
A11	0-8	Very dark gray (10YR3/1md) loam, subangular blocky (probably compacted due to asphalt and artifical fill above)
A12	8-33	Very dark gray loam, granular, numerous worm casts, charred plant remains, white saprolitic clasts
A13	33-36	Dark gray (10YR4/1d) loam in very dark gray loam matrix, granular
IIA31	36-51	Very dark gray loam, subangular blocky, numerous white saprolitic clasts
IIA32	51-71	As above with even more white saprolitic clasts increasing in size with depth
IIA33	71-92	Gray (10YR5/1d) to very dark gray loam, subangular blocky, numerous coarse white saprolitic clasts
IIA34	92-118	As above with light gray (10YR7/2d) mottles
IIIC1	118-168	Light gray (10YR7/2d, 6/2m) coarse sandy loam, trace of gravel, massive, numerous white saprolitic clasts
		Soil Profile D (1.5 m north of the main fault break)
A11	0-5	Very dark gray (10YR3/1md) loam, granular, (probably compacted due to asphalt and artificial fill above)
A12	5-35	Very dark gray loam, granular, numerous worm casts, charred plant remains, white saprolitic clasts
A13	35-40	Very dark gray loam, granular, some white silt
IIA31	40-57	Dark gray (10YR4/1d) loam, subangular blocky, numerous white saprolitic clasts larger than those above
IIA32	57-76	As above with even more white saprolitic clasts increasing in size with depth
IIA33	76-86	Dark gray loam, subangular blocky, numerous coarse white saprolitic clasts larger than those above
IIIB2g	86-121	Light gray (10YR7/2d) to olive (5Y5/3m) clay loam, subangular blocky, white saprolitic clasts in upper 5 cm
IIIB3g	121-126	Light gray (10YR7/2d) to light brownish gray (2.5YR7/2m) loam, subangular blocky
IVC1	126-171	Light gray (2.5YR2/2d) to light brownish-gray (2.5YR6/2m) loam, massive

*North end of trench above main break along Raymond fault. Cumulic Haplaquoll (Soil Survey Staff, 1975), August 23, 1977, at 34° 7.62' latitude, 118° 5.80' longitude and 183 m elevation. Profile overlain by 11 cm of coarse saprolitic gravel fill and 7 cm of asphalt which covers the driveway. Mediterranean climate. Parent material is Holocene alluvium alongside a former depression or sag pond. Slope 2%, aspect, south. Drainage poor, but no water in trench dug to 3 m. Soil moist throughout. Dioritic saprolites common (up to 15 cm diameter), increasing in size and number with depth. pH neutral to moderately acid.

Table 2.   Particle size distribution of profiles A and D from trench A-A' at San Marino High School

CDMG No. (/77)	Depth, cm	Horizon	Sand 2-0.05	Total Silt 0.05- 0.002	Clay <0.002	Very Coarse 2-1	Coarse 1-0.5	Medium 0.5- 0.25	Fine 0.25- 0.1	Very Fine 0.1- 0.05	Coarse 0.05- 0.02	Medium 0.02- 0.005	Fine 0.005- 0.002	Gravel >2 mm, % of whole sample

PROFILE A

332	0-8	A11	37.9	43.9	18.2	3.2	3.8	6.8	12.2	11.9	18.1	17.8	8.0	0.8
333	18-23	A12	34.2	45.7	20.1	2.0	3.0	5.4	12.8	11.0	17.8	16.2	11.7	0.0
334	33-36	A13	40.0	42.5	17.5	2.0	3.0	5.5	15.0	14.5	18.7	16.1	7.7	0.0
335	41-46	IIA31	40.3	40.7	19.0	3.0	3.0	6.8	14.2	13.3	19.2	15.5	6.0	0.0
336	56-61	IIA32	44.3	34.1	21.6	3.5	3.7	8.0	16.8	12.3	17.5	12.9	3.7	0.0
337	66-71	IIA32	43.0	35.5	21.5	3.2	3.1	7.9	16.8	12.0	17.0	14.2	4.3	0.0
338	71-76	IIA33	44.0	34.7	21.3	3.6	3.4	7.8	16.4	12.8	17.8	12.6	4.3	0.0
339	81-87	IIA33	44.7	34.2	21.1	3.5	3.7	7.8	17.0	12.7	16.5	13.6	4.1	0.0
340	97-102	IIA34	46.8	34.2	19.0	4.2	5.0	8.5	15.5	13.6	17.2	11.4	5.6	0.0
341	107-112	IIA34	44.0	35.9	20.1	3.8	4.0	8.2	16.6	11.4	17.6	13.4	4.9	0.8
342	118-123	IIIC1	59.1	28.7	12.2	7.6	8.4	11.2	18.8	13.1	14.8	10.2	3.7	2.3

PROFILE D

343	0-5	A11	41.7	42.0	16.3	4.3	4.9	8.3	13.3	10.9	14.4	17.9	9.7	0.8
344	15-20	A12	38.0	41.2	20.8	3.5	2.3	7.0	14.0	11.2	16.2	15.8	9.2	0.0
345	35-40	A13	45.0	38.0	17.0	3.2	3.8	6.7	16.1	15.2	16.0	15.0	7.0	0.0
346	46-51	IIA31	46.0	38.5	15.5	3.2	4.0	8.0	16.4	14.4	17.6	14.2	6.7	0.0
347	61-66	IIA32	45.0	35.0	20.0	3.0	3.8	7.5	16.7	14.0	18.7	10.5	5.8	0.0
348	71-76	IIA32	44.0	34.0	22.0	3.0	3.2	7.6	16.2	14.0	16.1	12.9	5.0	0.0
349	76-81	IIA33	44.0	33.8	22.2	3.0	3.8	7.0	15.3	14.9	16.0	13.8	4.0	0.0
350	86-91	IIIB22g	36.2	34.6	29.2	2.5	2.4	6.1	14.0	11.2	17.0	13.8	3.8	0.8
351	101-106	IIIB22g	32.9	37.9	29.2	2.0	1.8	4.4	13.1	11.6	17.4	12.7	7.8	0.0
352	116-121	IIIB23g	31.7	41.1	27.2	2.0	2.0	4.3	10.5	12.9	17.2	16.8	7.1	0.0
353	121-126	IIIB3g	32.8	41.7	25.5	2.0	1.7	4.1	13.0	12.0	15.2	18.3	8.2	0.0
354	126-131	IVC1	42.2	40.6	17.2	2.0	2.2	6.0	17.6	14.4	21.8	12.3	6.5	0.0
355	131-136	IVC1	44.0	39.5	16.5	2.0	2.0	5.2	18.1	16.7	20.0	13.9	5.6	0.0

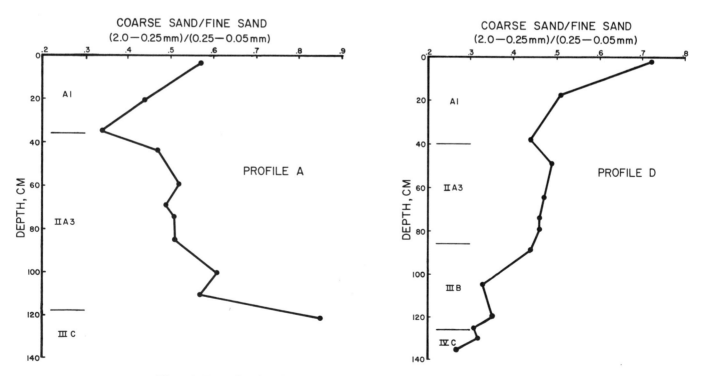

Figure 5. Depth functions for coarse sand to fine sand ratios in profiles A and D.

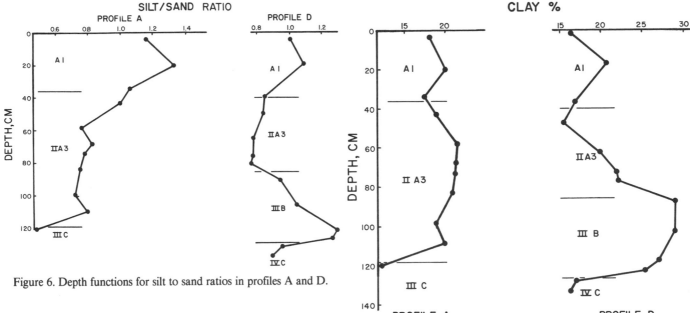

Figure 6. Depth functions for silt to sand ratios in profiles A and D.

Figure 7. Depth functions for clay contents in profiles A and D.

section. At least three slightly different alluvial units comprised the initial materials from which profile A formed, and at least four comprised the materials in profile D. Only one of these depositional units (now recognized as the Al horizon) extends uninterrupted across this strand of the fault. Conclusions about soil formation must take these heterogeneities into account.

PEDOGENESIS

Organic Matter Accumulation

As indicated by the "A" horizon designation in the description (Table 1), both profiles A and D have high accumulations of organic matter. Thick A horizons are defined by very dark gray to dark gray colors that extend to a depth of 118 cm in profile A and to 86 cm in profile D. In addition, small bits of charred or carbonized plant remains occur in the A12 horizons, but are absent below 35 cm. This evidence, combined with the particle-size data already presented, supports the "cumulic" designation in the soil's classification.

Carbon Dates

In well-drained soils of arid regions, organic matter is destroyed within a few centuries of its production, and carbon dates on what little soil organic matter remains are seldom useful. But in soils in which soil organic matter is not readily oxidized, carbon dates may be useful even though new carbon from modern plant remains is continually being mixed with the old carbon in the soil. Holocene prairie soils, for example, may have carbon dates that increase from about 500 B.P. at the surface to 5,000 B.P. at depths greater than 1 m (Yaalon, 1971, p. 83). As a rule, the more isolated the soil carbon is from modern influences, the older the radiocarbon date.

In profile D, soil organic matter from the 66 to 71 cm interval of the IIA32 horizon had a ^{14}C age of $5,360\pm300$ B.P. (Fig. 4), which confirms that this soil environment is conducive to the long-term preservation of organic matter. A charcoal sample from the 160-cm depth in the IIIC1 horizon of profile A gave a ^{14}C age of $10,100\pm300$ B.P. A sample of carbonaceous silt from the west wall at the 210-cm depth gave an age of $10,600\pm60$ B.P. Because it is surrounded by soil types not conducive to carbon accumulation (Eckmann and Zinn, 1917), the soil of the sag pond must have been poorly drained during most of its development.

Clay Accumulation

The curve that plots percent of clay against depth for well-developed Alfisols or Ultisols normally displays a noticeable bulge (the argillic B horizon) that marks the zone of clay accumulation. Although there is no confirming evidence for clay translocation (Table 1), profile D exhibits a clay bulge between 86 and 126 cm (Fig. 7). Curiously, profile A does not.

The existence of a B horizon on only one side of a dip-slip fault has two possible interpretations with regard to the recency of tectonic movement. If it can be shown that the B horizon has been eroded from the upthrown side of the fault, it is evidence for tectonic movement postdating development of the B horizon (Swan and Hanson, 1977; Borchardt, Rice, and Taylor, 1980; Borchardt, Taylor, and Rice, 1980; Borchardt and others, 1982). On the other hand, faults often bring together contrasting rock types. Soils developing in different parent materials do not necessarily produce B horizons at the same rate. This is especially true of young deposits, in which the initial clay content is a primary determinant of B-horizon development (Borchardt, Rice, and Taylor, 1980, p. 5). When clay is present in the initial materials, well-drained soils of arid regions may form argillic B horizons within 1,000 to 4,000 years (Gile, 1975, p. 349). When clay is absent in the initial materials, well-drained soils of arid regions sometimes may require 30,000 years to form argillic horizons (Marchand, 1977; Harden and Marchand, 1977).

We found no evidence of an erosive episode that could have removed a pre-existing B horizon from profile A. The differences between clay contents in profiles A and D could be a result of textural and, possibly, mineralogical differences in the initial materials on opposite sides of the fault.

Clay Mineralogy

Earlier we speculated that the parent material in this trench, granodioritic alluvium from the San Gabriel Mountains, may have obtained its clay fraction from sediments eroded from other soils in the Raymond Basin. Those soils, developed in well-drained older alluvium, are high in kaolinite (Buwalda, 1940, p. 16). There was, however, only a trace of kaolinite in any of our clay samples. Even though kaolinite does not form in poorly drained soils, it remains relatively stable under such conditions. Its scarcity in this trench indicates that the alluvial parent material was free of clay inherited from well-drained soils of the region. Because unweathered alluvial deposits from the San Gabriel Mountains usually contain little, if any, clay-size material (Crook and others, 1978, pp. 14, 22, and 26), the 12 to 29 percent clay that we found in the A, B, and C horizons from this site (Table 2) must have formed in situ.

This hypothesis was supported by the x-ray diffraction evidence that shows that the clay-size fractions in this soil are overwhelmingly dominated by well-crystallized minerals of the smectite group (Figs. 8 and 9). As outlined below, smectites are likely products of soil development under the poor drainage conditions existing at this site. The intensity of smectite peaks within the clay fraction increases dramatically with depth in the A1 horizon (Fig. 8), in keeping with the hypothesized young age of the alluvial unit from which it formed. Smectite peaks from samples of the A12 horizon on either side of the fault are nearly identical (Fig. 9), thus supporting the hypothesized deposition of the A1 horizon as a distinct alluvial stratum.

Within the smectite group, two minerals, beidellite and montmorillonite, can be distinguished with special techniques (Fig. 9). When solvated with glycerol vapor, beidellite gives a 14Å x-ray diffraction peak, while montmorillonite gives an 18Å peak (Harward and Brindley, 1964; Borchardt, 1977b, p. 296). Following these criteria, the A horizons in this soil all contain beidellite, whereas the B and C horizons in profile D contain only montmorillonite. Such information on smectite types can help determine the genesis of the horizons in which they exist. For example, it is unlikely that the B horizon in profile D would be without beidellite if it had formed via clay translocation from above. This interpretation is consistent with the notable absence

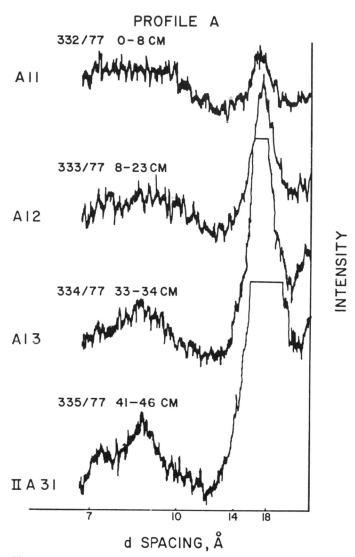

Figure 8. X-ray diffraction peaks illustrating that smectite contents increase with depth in the upper portion of profile A (Mg saturation, glycol treatment with full extent of lower peaks not shown).

of argillans or "clay skins" in the B horizon (Table 1). The B horizon is a cambic, not an argillic, horizon.

One mechanism for the formation of the smectite in situ is called clay transformation: the chemical and physical change of primary layer silicates, such as mica and chlorite, to other layer silicates, such as vermiculite, smectite, pedogenic chlorite, and kaolinite. In granodioritic and dioritic alluvium, the most likely mineral candidate for transformation to smectite is biotite. This is clearly visible in the saprolitic clasts and coarse fractions found in sediments underlying the soil profile. However, neither biotite nor the intermediate mineral in this mechanism, vermiculite, was detected in x-ray diffraction patterns of the clay fractions. Either this mechanism is not significant in this soil or the biotite → vermiculite → smectite transformation (Jackson, 1965; Borchardt and others, 1966) has gone to completion. This seems unlikely in view of the absence of mica or vermiculite peaks in x-ray patterns

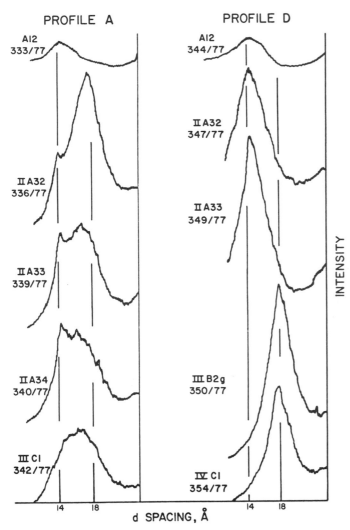

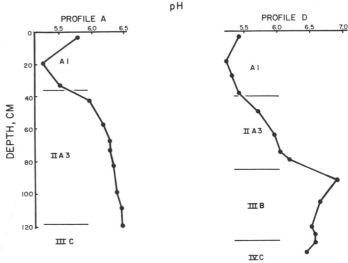

Figure 10. Depth functions for pH in profiles A and D.

Figure 9. X-ray diffraction peaks illustrating beidellite-montmorillonite mixtures in profile A and beidellite overlying montmorillonite in profile D (Mg saturation, glycerol vapor treatment).

of the A11, the youngest and apparently the least developed of the horizons (Fig. 8).

Another mechanism for the formation of smectite in situ is clay synthesis or neogenesis, the precipitation or crystallization of layer silicates from soil solution. It is well known that smectites, particularly montmorillonite, form in poorly drained soils—environments that concentrate the necessary soluble silica and bases and have pH levels close to neutrality (Jackson, 1965; Borchardt, 1977b, p. 306).

The depth function for pH in profile A is similar to that found in many Holocene soils (Fig. 10). In the uppermost horizon the pH is relatively high owing to base compounds of calcium and manganese brought from depth and recycled to the soil by root decay. With increasing depth, however, pH decreases as a result of downward percolation of meteoric water containing humic, fulvic, and carbonic acid.

The lower portion of profile D has a pH depth function that

reflects its unique clay mineralogy. As in profile A, the beidellitic horizons have pH values less than 6.5. The montmorillonitic horizons, however, have pH values greater than 6.5 (Figs. 9 and 10). The pH jumps from 6.2 to 7.0 precisely at the boundary between the IIA3 and IIIB horizons, where the clay mineralogy changes from beidellite to montmorillonite. There is likewise a large increase in clay and total smectite content across this boundary (Figs. 7 and 11). Soils with equilibrium pH values greater than 6.7 generally are fully base saturated (e.g., Borchardt and others, 1968, p. 400), that is, the cation exchange complex contains only the basic ions such as calcium, magnesium, potassium, and sodium, rather than the acidic ions of the aluminum or hydronium complex (Jackson, 1963). Our data thus indicate that the acidic ions are necessary for the introduction of tetrahedral aluminum into the 2:1 layer silicate structure during beidellite neogenesis.

Total iron contents of the <420 μm fractions indirectly provide a clue to the neogenesis of montmorillonite. Iron contents in the A1 horizon decrease with depth on both sides of the fault but increase sharply at the contact with the second alluvial unit (Fig. 12). This is further evidence that the A1 horizon developed in a distinct alluvial unit. The montmorillonitic B horizon has the highest total iron, total clay (Fig. 7), and total smectite contents (Fig. 11), in addition to the highest pH. The iron-containing minerals in the parent material are mostly hornblende and biotite (Crook and others, 1978, p. 10). In other Holocene soils, hornblende disappears from silt fractions as montmorillonite appears in clay fractions (Glenn and others, 1960). In our study, the silt/sand ratios in the B horizon decrease (Fig. 6) as the smectite and iron contents increase (Figs. 11 and 12). As mentioned, it is possible that the silt/sand relationship in the B horizon is an original artifact of alluvial deposition, but it is also likely that the chemical dissolution of silt-size hornblende furnished the high base saturation that accelerated the neogenesis of montmorillonite in the upper portion of that unit.

RELATIVE AMOUNT OF SMECTITE IN SOIL

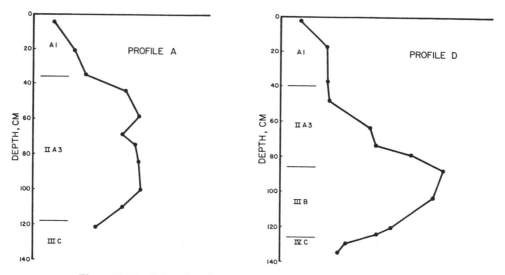

Figure 11. Depth functions for relative smectite contents of profiles A and D.

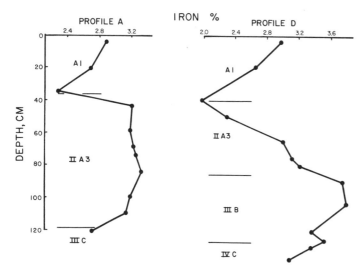

Figure 12. Depth functions for total iron in the <420 μm fraction of profiles A and D.

TABLE 3. SIMAN COEFFICIENTS FOR COMPARISONS BETWEEN A TECTONICALLY PRODUCED SOIL TONGUE AND SOIL FROM PROFILES A AND D

Horizon	Profile A	Profile D
A11	0.74*	0.81
A12	0.78	0.81
A13	0.83	0.88
IIA31	0.90	0.90
IIA32	0.93	0.93
IIA33	0.94	0.93
IIIB2g	na	0.74
IIIC1	0.74	na
IVC1	na	0.82

*A coefficient of 1.00 reflects perfect similiarity, while a coefficient of 0 reflects perfect dissimilarity. These multivariate comparisons include all data from partical size fractions in Table 2, pH, total CA (data not shown), total Fe, and the ratios of Ca/Fe, verycoarse sand/very fine sand, and silt/sand. For a discussion of the SIMAN coefficient, see Borchardt (1972; 1974). The soil tongue was sampled between 102 and 127 cm (sample 359/77). na = not applicable.

TECTONISM

A distinctive soil tongue extends from 62 to 137 cm within the tectonically disturbed area between profiles A and D on the trench wall opposite the one in Fig. 4. The origin and age of this feature have implications for the date of the most recent movement on this strand of the Raymond fault. For example, if the tongue contained soil similar to the modern surface, we would consider the tongue to have been produced after the deposition and formation of the A1 horizon. To this effect, we used similarity analysis (Borchardt, 1972; 1974; Borchardt, Rice and Taylor, 1980, p. 32) to compare soil from the tongue with the soil samples from profiles A and D (Table 3). The soil from the tongue was more similar to the soil in the IIA33 horizons than it was to the soil in the A1, B, or C horizons (Table 3). Therefore, we conclude that the soil in the tongue is not an infilling derived from the present surface. The pH throughout the tongue was 6.0 and the iron content was 2.8 percent, values similar to those found at the 60-cm level in profile D (Figs. 10, 12). The pH and the iron content vary widely throughout profiles A and D, and thus we conclude that the soil tongue has not weathered significantly since emplacement.

Soil carbon in the bottom of the tongue gave a date of 6,060±110 B.P. (John Tinsley, personal commun., 1980), whereas soil carbon in the horizon from which it may have fallen during tectonic movement gave a date of 5,360±300 B.P. (Fig. 4). By assuming that the carbon in the soil tongue was not subject to additions of modern carbon after its emplacement, we can speculate that the difference between these two dates, 700±320 years, is up to half the time elapsed (1,400 years) since the event that produced the soil tongue. The unfaulted alluvial unit (the present-day A1 horizon) that buried the soil tongue and the underlying IIA3 horizon may correlate with the Qal_1 geologic unit that Crook and others (1978, p. 30) consider to be about 1,000 years old.

CONCLUSIONS

We conclude that:

1. Soil profiles A and D developed from six slightly different alluvial beds, two of which are coarse-grained and unique to the south side and three of which are fine-grained and unique to the north side of a strand of the Raymond fault.

2. Clay fractions from the site contained only a trace of kaolinite, indicating that the parent material, granodiorite, was relatively free of clay minerals inherited from upland soils. Soil development began here about 10,000 years ago. Poor drainage was responsible for the neogenetic synthesis of minerals of the smectite group. Beidellite formed in horizons containing exchangeable aluminum (pH less than 6.5), while montmorillonite formed in horizons saturated with exchangeable bases (pH greater than 6.5) probably released from especially high concentrations of silt-size hornblende.

3. Only the A1 horizon (the upper 40 cm of the soil profile) overlies both soil profiles; its deposition occurred shortly after the formation of a soil tongue produced by tectonic movement. The soil in the tongue has weathered little since its emplacement less than 1,400 years ago.

4. The relative degree of soil development within this Cumulic Haplaquoll is halfway between that of the Hanford soils of the San Joaquin Valley (Harden and Marchand, 1977) and the Concepcion soils of the coast (Shlemon, 1978; Borchardt and others, 1982).

5. This study confirms that this strand of the Raymond fault poses a significant seismic risk to structures built upon it.

ACKNOWLEDGMENTS

This paper is dedicated to the memory of Denis E. Marchand, one of the first geologists to recognize the significance of pedogenesis for evaluating fault activity. We thank Glenn A. Brown, Chief Geologist for LeRoy Crandall and Associates, Los Angeles, California, for providing information about and access to the trench at San Marino High School; David Smith, graduate student at the University of Southern California, for field assistance; Marc Loza, Chuck Smith, and Hasu Majmundar of the California Division of Mines and Geology Geochemical Section for laboratory analyses; and Wilma Ashby, Venice Huffman, Esther NeSmith, Delores Abraham, and Sue Torres for typing the manuscript.

This work was sponsored by the National Earthquake Hazards Reduction Program of the U.S. Geological Survey under Contract No. 14-08-0001-15858 and Grant No. 14-08-0001-G-510. The views and conclusions contained in this article are those of the authors and should not be interpreted as necessarily representing the official policies, either expressed or implied, of the U.S. Government or the State of California.

REFERENCES CITED

Borchardt, G. A., 1972, Geochemical similarity analysis: Geocom Programs, v. 6, p. 1–31.
—— 1974, The SIMAN coefficient for similarity analysis: Bulletin of the Classification Society, v. 3, p. 2–8.
—— 1977a, Clay mineralogy and slope stability: California Division of Mines and Geology Special Report 133, 15 p.
—— 1977b, Montmorillonite and other smectite minerals, *in* Dixon, J. B., and Weed, S. B., eds., Minerals in soil environments: Soil Science Society of America, Madison, p. 293–330.
—— 1977c, Physical testing of soils for urban development: California Geology, v. 30, p. 254–258.
Borchardt, G. A., Hole, F. D., and Jackson, M. L., 1968, Genesis of layer silicates in representative soils in a glacial landscape of southeastern Wisconsin: Soil Science Society of America Proceedings, v. 32, p. 399–403.
Borchardt, G. A., Jackson, M. L., and Hole, F. D., 1966, Expansible layer silicate genesis in soils depicted in mica pseudomorphs: Proceedings, International Clay Conference, Jerusalem, Volume 1, p. 175–185.
Borchardt, G. A., Rice, S., and Taylor, G., 1980, Paleosols overlying the Foothills fault system near Auburn, California: California Division of Mines and Geology Special Report 149, 38 p.
Borchardt, G. A., Rice, S., and Treiman, J., 1982, Mudflow deposits and horizonation in Holocene soils near Point Conception, California [abs.]: Geological Society of America Abstracts with Programs, Volume 14, p. 150.
Borchardt, G. A., Taylor, G., and Rice, S., 1980, Fault features in soils of the Mehrten Formation, Auburn damsite, California: California Division of Mines and Geology Special Report 141, 45 p.
Borchardt, G. A., and Theisen, A. A., 1971, Rapid x-ray spectrographic determination of major element homogeneity in Mazama pumice: Soil Science and Plant Analysis, v. 2, p. 11–16.
Brown, G., and Farrow, R., 1956, Introduction of glycerol into flake aggregates by vapour pressure: Clay Minerals Bulletin, v. 3, p. 44–45.
Bryant, W. A., 1978, The Raymond Hill fault: An urban geological investigation: California Geology, v. 31, p. 127–142.
Buntley, G. J., Daniels, R. B., Gamble, E. E., and Brown, W. T., 1977, Fragipan horizons in soils of the Memphis-Loring-Grenada sequence in west Tennessee: Soil Science Society of America Journal, v. 41, p. 400–407.
Buwalda, J. P., 1940, Geology of the Raymond basin: Unpublished report for the Pasadena Water Department, 131 p.
Chu, T. Y., and Davidson, D. T., 1953, Simplified air jet dispersion apparatus for mechanical analysis of soils: Proceedings of the Highway Research Board, v. 32, p. 541–547.
Crook, R., Jr., Allen, C. R., Kamb, B., Payne, C. M., and Proctor, R. J., 1978,

Quaternary geology and seismic hazard of the Sierra Madre and associated faults, western San Gabriel Mountains, California: Contribution No. 3191, Division of Geological and Planetary Sciences, California Institute of Technology, 117 p.

Day, P. R., 1965, Particle fractionation and particle-size analysis, *in* Black, C. A., ed., Methods of soil analysis, Part 1: Madison, Soil Science Society of America, p. 545–567.

Eckmann, E. C., and Zinn, C. J., 1917, Soil survey of the Pasadena area, California: U.S. Department of Agriculture Report, 56 p.

Flanagan, F. J., 1973, 1972 values for international geochemical reference samples: Geochimica et Cosmochimica Acta, v. 37, p. 1189–1200.

Gile, L. H., 1975, Holocene soils and soil-geomorphic relations in an arid region of southern New Mexico: Quaternary Research, v. 5, p. 321–360.

Glenn, R. C., Jackson, M. L., Hole, F. D., and Lee, G. B., 1960, Chemical weathering of layer silicate clays in loess-derived Tama silt loam of southwestern Wisconsin: Clays and Clay Minerals, v. 8, p. 63–83.

Harden, J. W., and Marchand, D. E., 1977, The soil chronosequence of the Merced River area [California], *in* Singer, M. J., ed., Soil development, geomorphology, and Cenozoic history of the northeastern San Joaquin Valley and adjacent area, California: Joint Field Session, American Society for Agronomy, Soil Science Society of America, and Geological Society of America, Guidebook, Davis, University of California Press, p. 22–38.

Harward, M. E., and Brindley, G. W., 1964, Swelling properties of synthetic smectites in relation to lattice substitutions: Clays and Clay Minerals, v. 13, p. 209–222.

Hill, R. L., Sprotte, E. C., Bennett, J. H., Chapman, R. H., Chase, G. W., Real, C. R., and Borchardt, G. A., 1977, Santa Monica-Raymond Hill fault zone study, Los Angeles County, California: California Division of Mines and Geology unpublished annual technical report for the U.S. Geological Survey National Earthquake Hazards Reduction Program.

Hill, R. L., Sprotte, E. C., Chapman, R. H., Chase, G. W., Bennett, J. H., Real, C. R., Borchardt, G. A., and Weber, F. H., 1978, Earthquake hazards associated with faults in the greater Los Angeles metropolitan area, Los Angeles County, California, including faults in the Santa Monica-Raymond Hill, Verdugo-Eagle Rock, and Benedict Canyon fault zones: California Division of Mines and Geology unpublished annual technical report for the U.S. Geological Survey National Earthquake Hazards Reduction Program.

Jackson, M. L., 1963, Aluminum bonding in soils: A unifying principle in soil science: Soil Science Society of America Proceedings, v. 27, p. 1–10.

——1965, Clay transformations in soil genesis during the Quaternary: Soil Science, v. 99, p. 15–22.

Marchand, D. E., 1977, The Cenozoic history of the San Joaquin Valley and adjacent Sierra Nevada as inferred from the geology and soils of the eastern San Joaquin Valley, *in* Singer, M. J., ed., Soil development, geomorphology, and Cenozoic history of the northeastern San Joaquin Valley and adjacent areas, California: Joint Field Session, American Society for Agronomy, Soil Science Society of America, and Geological Society of America, Guidebook, Davis, University of California Press, p. 39–50.

Shlemon, R. J., 1978, Soil-stratigraphy of late Pleistocene-Holocene deposits, proposed LNG site, Point Conception, California: Appendix A in Dames & Moore, addendum report, Geological investigations, proposed LNG terminal, Point Conception, California. Unpublished consulting report, Job No. 00011-168-02 for Western LNG Associates, July 12, 1978, 17 p.

Soil Survey Staff, 1951, Soil survey manual: U.S. Department of Agriculture Handbook No. 18, Washington, D.C., U.S. Government Printing Office, 503 p.

——1975, Soil taxonomy: A basic system of soil classification for making and interpreting soil surveys: U.S. Department of Agriculture, Soil Conservation Service, Agricultural Handbook No. 436: Washington, D.C., U.S. Government Printing Office, 754 p.

Swan, F. H., III, and Hanson, K. L., 1977, Quaternary geology and age dating. Earthquake evaluation studies of the Auburn dam area: Woodward-Clyde Consultants Report for U.S. Bureau of Reclamation, volume 4.

Theisen, A. A., and Harward, M. E., 1962, A paste method for preparation of slides for clay mineral identification by x-ray diffraction: Soil Science Society of America Proceedings, v. 26, p. 90–91.

Yaalon, D. H., 1971, Paleopedology: Origin, nature, and dating of paleosols: Jerusalem, International Society of Soil Science and Israel University Press, 350 p.

MANUSCRIPT ACCEPTED BY THE SOCIETY JANUARY 12, 1985

Geological Society of America
Special Paper 203
1985

The Quaternary alluvial sequence of the Antelope Valley, California

Daniel J. Ponti

U.S. Geological Survey
345 Middlefield Road
Menlo Park, California 94025
and
Department of Geological Sciences
University of Colorado
Boulder, Colorado 80309

ABSTRACT

Arkosic alluvium derived from the Transverse Ranges and Tehachapi Mountains and deposited in the closed basin of the Antelope Valley, California, can be subdivided into eight mappable units. The six oldest units range in age from less than 0.5 m.y. through early Holocene and form widespread fill terraces and fans that extend over several thousand km² of the valley floor. In contrast, the two youngest units, which consist of late Holocene and modern floodplain and channel sediments, are thin and of limited areal extent. The youngest of the six older deposits can be physically traced into nearly all drainages along both mountain ranges, which suggests nearly synchronous deposition across the region during latest Pleistocene to early Holocene time. These deposits appear to interfinger with and overlie pluvial lake deposits. The five older deposits can be recognized and correlated based on a combination of several factors such as similarity of soils, in situ shear-wave velocities, and degree of primary surface dissection. This evidence suggests that principal alluviation was regionally synchronous during the middle and late Pleistocene as well.

The existence of a pervasive and uniform sequence of apparently synchronous deposits throughout the region suggests that major aggradational and degradational episodes are controlled by regional climatic fluctuations. The existence of buried soils that require long periods of time for their formation further supports a climatic control of alluviation by suggesting that principal alluvial deposition is rapid and episodic. Finally, sedimentological evidence that indicates a change in hydrologic conditions during principal alluviation and an apparent correlation among major units in the Antelope Valley to those in the eastern San Joaquin Valley strongly suggest that aggradational episodes occur during major glacial to interglacial climatic transitions.

INTRODUCTION

The Antelope Valley, a sediment-filled, closed basin of about 4,000 km², lies between the seismically active San Andreas and Garlock fault zones and forms the westernmost "wedge" of the Mojave Desert geologic province of southern California (Fig. 1). Elevations within the region range from about 700 m on the basin floor to over 1,200 m in the foothills. The climate is semi-arid to arid; mean annual temperatures range from about 15° to 20°C, and mean annual rainfall averages 13 to 46 cm (Woodruff and others, 1970). Almost all of the rain falls between November and April, which suggests that the soil-moisture regime is xeric. Winds are generally out of the north and west and are light, except during the spring when gale force winds are common. Vegetation on the basin floor consists of saltgrass and sage. With increasing elevation, the foothills are mantled with

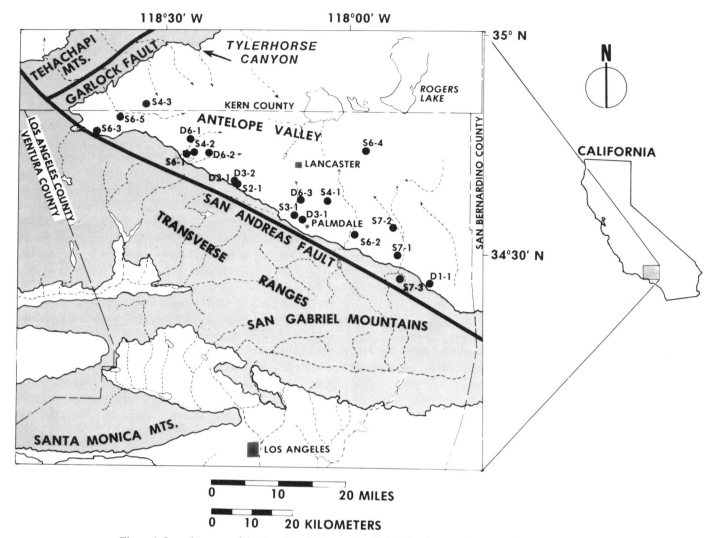

Figure 1. Location map of the Antelope Valley region. Labeled points are the approximate locations of the soil-sample sites in Table 1.

prairie grass and sage interspersed with stands of joshua trees, juniper, and pine (Woodruff and others, 1970).

The Quaternary basin sediments are derived from the Transverse Ranges and Tehachapi Mountains, which border the valley on the southwest and northwest, respectively. Prior to Pleistocene time, however, the drainage networks were considerably different because much of the Mojave Desert region was probably an upland area with a regional paleoslope to the west (Woodburne, 1975; Woodburne and Golz, 1972; Dibblee, 1967). During the late Tertiary, streams headed in the interior of the Mojave Desert province and flowed to the Pacific Ocean, causing continental deposition in basins aligned approximately east-west. In contrast, modern streams head in the Transverse Ranges and Tehachapi Mountains and flow into Rogers and Rosamond playas. This change in the drainage pattern apparently resulted from extensive uplift of the San Gabriel Mountains. Fossil evidence and sediment sources suggest that this uplift and the resultant depositional changes probably started no earlier than 2

m.y. ago and perhaps began as late as 1 m.y. ago (C. Repenning, written commun., 1982; Woodburne, 1975). As a result, much of the 2,000 m of sediment presently in the Antelope Valley basin may have been deposited during the Quaternary.

Noble (1953, 1954), Dibblee (1967), and Woodburne and Golz (1972) mapped the bedrock and recognized that several bedrock formations are probably Pleistocene in age. The Pleistocene Harold Formation is the oldest formation recognized by Noble (1953, 1954) and Dibblee (1967) that contains granitic clasts derived predominantly from the San Gabriel Mountains. Later work by Woodburne and Golz (1972) and Woodburne (1975) suggests that the Crowder Formation and the "western facies" (Woodburne, 1975) portion of the Punchbowl Formation may extend into the Pleistocene as well. These lower Quaternary sediments, which record the rise of the San Gabriel "Arch" (Woodburne, 1975), are extremely deformed and are exposed only near the San Andreas rift zone and adjacent to small east-west trending thrust faults and folds that are common along the northern margin of the Transverse Ranges. Comparable sediments are not exposed along the south flank of the Tehachapi Mountains, except in the core of one anticline (Ponti and others, 1981). These older sediments are exposed only along the north flank of the Transverse Ranges and are not generally exposed along the south flank of the Tehachapi Mountains because there are different styles of tectonism extant along each mountain range. East-west trending reverse faults and folds are present locally along much of the north flank of the Transverse Ranges. These structures have significantly uplifted, warped, and exposed the older sediments along the range-front margins. Reverse faults and folds, however, are not as prevalent along the south flank of the Tehachapi Mountains. Instead, this mountain front is characterized by related sets of northwest- and southwest-trending strike-slip faults. Strike-slip deformation along the south flank of the Tehachapi Mountains does not produce the amount of basin-margin upwarping and erosion that exists along the Transverse Ranges margin.

Alluvial deposits that overlie the lower Quaternary sediments were mapped by Noble (1953, 1954) and Dibblee (1967) as the Shoemaker Gravel, Nadeau Gravel, old terraces, and old and young alluvium. These deposits, which are composed entirely of detritus derived from the Transverse Ranges and Tehachapi Mountains, form fans and terraces whose primary surfaces have been completely or partially preserved. The oldest of these deposits, the Shoemaker and Nadeau Gravels, are distinguished from each other by the relative amount of clasts derived from the Pelona Schist; the Nadeau Gravel contains a high amount of schist whereas the Shoemaker Gravel does not. Despite the variations in source lithology, these two units are probably stratigraphically equivalent. Clast lithology is a useful tool for correlating the Shoemaker and Nadeau Gravels across the San Andreas fault, but it is not very useful for interbasin correlations. The old terraces and old and young alluvium mapped by Noble (1953, 1954) and Dibblee (1967) were differentiated primarily on the basis of their morphologic features. However, Noble and Dibblee made no attempt to establish a detailed stratigraphic sequence within these deposits.

This paper summarizes the results of recent mapping of post-Harold Formation alluvial sediments (Ponti and others, 1981; Ponti and Burke, 1980; Ponti, 1980). The purposes of these studies were to better define the stratigraphy of the alluvial deposits and to improve the assessment of the timing and the degree of late Quaternary faulting and deformation (cf. Dibblee, 1967).

Six major upper Quaternary alluvial units are recognized and correlated throughout the region. These units are probably a result of five or six episodes of relatively rapid and extensive alluvial fan aggradation that postdate deposition of the Harold Formation. In addition, two alluvial units of limited areal extent and thickness are also recognized; they represent late Holocene aggradation in modern channels and floodplains. The widespread occurrence of this sequence of deposits and other evidence suggest that the principal episodes of alluvial deposition are roughly synchronous throughout the basin. Given that the styles of de-

formation along the margins of the Transverse Ranges and Tehachapi Mountains are different, but the Quaternary sediments are of equivalent age, a tectonic origin seems improbable. A more reasonable explanation is that deposition is related to some change in regional climate. If the regional climate has changed in response to global changes, it may be possible to correlate the alluvial units in the Antelope Valley with those in other areas.

THE QUATERNARY ALLUVIAL SEQUENCE OF THE ANTELOPE VALLEY

Mapping Criteria

The Antelope Valley alluvial deposits are defined and correlated on the basis of superposition, the degree of soil development on original surfaces, and geomorphic criteria such as relative topographic position and degree of dissection of original surfaces. The lithology of clasts within different deposits is not useful when attempting to distinguish between the deposits because most upper Quaternary deposits of the region are derived from the same or similar rocks.

Degree of soil development, which is time dependent, may be the most useful criterion for correlating the units because at least some primary surfaces of most deposits are preserved (see Richmond, 1962; Scott, 1963; Janda and Croft, 1967; Birkeland, 1974; Birkeland and others, 1979; Gile and others, 1981; and Tinsley and others, 1982, for examples of the use of soils in Quaternary stratigraphy). However, soil development is also a function of differences in climate, vegetation, relief, and parent material (Jenny, 1941; Birkeland, 1974). Within the Antelope Valley, variations among soils on coeval deposits are noted between the basin floor and the foothills where annual rainfall amounts can differ by as much as 33 cm, and vegetation types are observed to differ accordingly. In general, soils have redder colors and are more strongly developed where rainfall is greater. In addition, soils on apparently correlative deposits also vary on a regional basis as a result of differences in source-rock lithology. For the most part, the bedrock of both the Tehachapi Mountains and the Transverse Ranges is dominantly of granitic origin, ranging in composition from quartz diorite to quartz monzonite. However, small bodies of gneissose and schistose rocks are present locally, particularly near the town of Palmdale. Deposits derived from these rocks, which have a high percentage of amphibole and biotite, commonly have soils with redder colors and higher clay contents than do soils on coeval deposits derived from the granitic rocks. Lastly, variations in soil properties that result from differences in the texture of the alluvial material are also noted within the basin. Unweathered alluvial detritus (soil parent material) within about 15 km of the front of the Tehachapi Mountains and the Transverse Ranges is composed almost entirely of sand or gravelly sand, with less than 4 percent silt and clay. In more distal areas, parent-material textures are finer and range from sandy loam to silt loam (25 to 70 percent silt and clay, respectively). Argillic horizons formed in the fine-grained sediments are com-

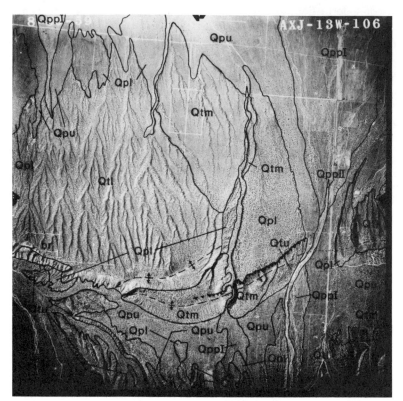

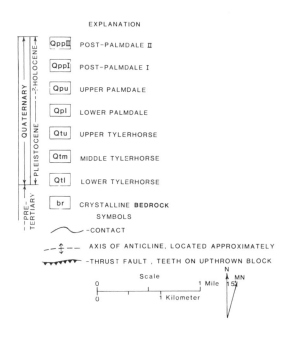

Figure 2. Photogeologic map showing general surface characteristics of the upper Quaternary sediments east of Holcomb Ridge in the Valyermo quadrangle along the north flank of the San Gabriel Mountains. The large area of lower and middle Tylerhorse sediments (center left) can be distinguished by the high density of entrenched streams. Upper Tylerhorse sediments in the southern portion of the photo are graded to a much lower level. The preserved braided stream-channel morphology on the surface of the post-Palmdale I deposit to the northeast is typical for that unit.

monly more strongly developed than argillic horizons formed in coeval, coarse-grained material.

Because the numerous factors of soil formation are interdependent (that is, parent-material textures fine basinward while rainfall decreases basinward), it is difficult to account for the effects of any given factor when comparing soils on correlative deposits. Given that soil profiles developed in apparently coeval deposits occasionally do differ, it becomes necessary to incorporate fan-distribution and fan-morphology characteristics to aid in correlation. Criteria commonly used include fan slope, height above present stream channel (topographic position), and fan surface dissection. Within the Antelope Valley, these morphologic characteristics, particularly fan-surface dissection, are often good indicators of relative age (Fig. 2); young fans have well preserved surfaces, whereas those of old deposits are highly dissected. Fan morphology also becomes a most important correlation tool in areas where soils are stripped or are buried by thin veneers of Holocene eolian sand. Where the soils are preserved, fan morphology can provide corroborative evidence for correlation; well-developed soils occur on old, dissected fans, and successively weaker soils occur on successively younger and less dissected fans. The distribution of the deposits also provides an aid to correlation; old fans and old deposits tend to be restricted to the base of the mountains, whereas younger deposits occur predominantly on the valley floor. Where the soils data and fan-morphology information appear to conflict, the discrepancy can usually be attributed to tectonic deformation of the sediments. In these areas, detailed correlation is sometimes difficult.

Description of Alluvial Deposits within the Antelope Valley

Tylerhorse Deposits. The three oldest major alluvial units in the Antelope Valley are informally called the upper, middle, and lower Tylerhorse deposits, after exposures near Tylerhorse Canyon in the foothills of the Tehachapi Mountains. As mapped, these sediments include the Shoemaker and Nadeau Gravels of Noble (1953, 1954), as well as finer grained sediments of similar age that were mapped by Noble (1953, 1954) and Dibblee (1967) as old terraces and old alluvium. In some places, these deposits are observed to unconformably overlie the Harold Formation.

Soils on the Tylerhorse deposits are the most highly developed alluvial soils in the basin. Thick (0.5 to >2 m), well developed argillic horizons are present in these soils. They exhibit subangular blocky to prismatic structure and are typically reddish brown (7.5YR to 5YR hues, with chromas up to 6). The sediments range from moderately consolidated to unconsolidated and form moderately to extensively dissected and tilted terraces and fans exposed along the foothills and in regions of active folding within the basin. On the basis of stratigraphic position, degree of soil development, and comparison with other alluvial sequences, these sediments are almost certainly Pleistocene in age, and are probably greater than 100,000 years old.

The lower Tylerhorse deposits, which are confined to the mountain fronts, are generally uplifted and are not obviously related to the modern drainage networks. Fan and terrace surfaces formed on these deposits are extensively dissected, and in many cases fewer than an estimated 10 percent of the original landforms are preserved. Surface soils are very well developed in this unit; they contain argillic horizons that can be more than 2 m thick and commonly contain greater than 30 percent clay. Soil colors in the argillic horizon are generally reddish brown (5YR hues), and individual clasts within the horizon are so weathered that they can be easily crushed with the fingers. Locally, the lower Tylerhorse unit exceeds 70 m in thickness.

Deposits of the middle Tylerhorse unit, which are locally inset into lower Tylerhorse deposits, are commonly exposed farther from the mountain front than are deposits of the lower unit, but are still primarily restricted to the foothills. These sediments are also uplifted, and many bear no relationship to the modern drainage systems. The topographic surfaces formed on deposits of the middle unit are much better preserved than those formed on the lower unit, with as much as 50 percent of the original landforms still extant. Soils formed on this unit are well developed, with argillic horizons that are commonly no thicker than 1.5 m and that usually contain no more than 30 percent clay. Soil colors in the argillic horizon range from light reddish brown to light brown (5YR to 7.5YR hues), and clast weathering is not as pronounced as in the lower unit. Exposed thicknesses of this unit exceed 30 m.

Upper Tylerhorse deposits were laid down as terraces in canyons cut into lower and middle Tylerhorse deposits, or as fans located farther out in the basin. Most deposits correlate with present drainage systems. Surfaces of the upper Tylerhorse deposits are moderately dissected, and generally more than 70 percent of the original landforms are preserved. Surface soils are well developed, with an argillic horizon usually less than 1 m thick and with less than 25 percent clay. The reddish brown to brown colors in the B horizon have 7.5YR hues and rarely are they as red as 5YR, except in areas where deposits are derived from a predominantly schistose source. Clasts with a high mafic mineral content (such as those from the Pelona Schist) are well-weathered and are easily broken apart by hand, but clasts of granitic composition are relatively fresh. Some exposures of the upper Tylerhorse unit exceed 20 m in thickness.

Palmdale Deposits. The next youngest units, informally called the lower, middle, and upper Palmdale deposits, cover most of the basin floor, but also occur along the margins of the basin as terraces within broad canyons cut into the Tylerhorse deposits. These sediments were previously mapped as young and old alluvium by Dibblee (1967) and Noble (1953, 1954). Palmdale deposits form broad alluvial fan and terrace surfaces that are moderately entrenched and, in some areas, tilted; they are closely associated with the modern drainage systems, and the profiles of the fan surfaces converge downstream with the modern channel profiles. These sediments are distinguishable from Tylerhorse deposits on the basis of the weakly developed soils formed on Palmdale deposits. Soils on Palmdale deposits, while exhibiting a good deal of variability, have only cambic and weak argillic horizons and lack the reddish brown hues and the well-developed argillic horizons characteristic of soils on Tylerhorse deposits. By stratigraphic position, degree of soil development, and comparison with other alluvial sequences, these sediments are estimated to range in age from late Pleistocene to early Holocene.

Lower Palmdale deposits form slightly uplifted and tilted terraces and fan surfaces that are commonly entrenched 5 m or more, but are only slightly eroded (about 90 percent of the surfaces are preserved). Soils on lower Palmdale sediments exhibit weak to moderately developed argillic horizons, generally less than 1 m thick, that have at maximum a moderate subangular blocky structure; clay content of the argillic horizon in sandy facies of these deposits rarely exceeds 15 percent. Soil colors range from pale to dark brown (10YR hues), and clasts of granitic composition remain relatively fresh, whereas schistose material is commonly weathered and incompetent. Thicknesses of this unit may commonly exceed 20 m.

Deposits mapped as middle Palmdale are always inset into lower Palmdale deposits and are topographically higher than adjacent upper Palmdale surfaces. Exposures of this unit do not reveal whether these deposits represent fill terraces or are strath surfaces cut into lower Palmdale deposits. Deposits of this unit exist in only a few locations and can be recognized only where they occur between an upper and lower Palmdale surface. The limited extent of this unit suggests that these deposits probably do not represent a major episode of aggradation. Soils on this unit are variable, but they do exhibit a degree of development intermediate between soils formed on the upper and lower units. Cambic horizons are common, but thin argillic horizons exist locally.

Upper Palmdale deposits are exposed over a large portion of the Antelope Valley basin. Deposits of this unit form low terraces in valleys along the foothills and grade into broad, smooth alluvial fans that blanket the basin floor. The surfaces of these deposits are generally flat-lying and undissected, but modern channels are incised at least 1 m. At the distal edges of the alluvial fans, in the vicinity of Rogers and Rosamond dry lakes, several exposures suggest that upper Palmdale deposits apparently interfinger with, and overlie, lacustrine sediments presumably deposited during the last major pluvial period (Fig. 3). Soils on the upper Palmdale deposits are weakly developed, with only an A and/or Cox profile. Soil colors range from pale to dark brown (10YR hues), and clasts are relatively fresh, although some are internally oxidized. Thicknesses of this unit locally exceed 12 m.

Post-Palmdale Deposits. Post-Palmdale deposits in this

UPPER PALMDALE ALLUVIUM

~1m

COVERED

CROSS-BEDDED COARSE SAND (LAKE BAR?)

LACUSTRINE (?) SILT

ALLUVIUM (?)

Figure 3. Exposure in a sand pit (Rosamond Lake quadrangle; NE ¼, NW ¼, Sec. 29, T8N, R11W) near the former shoreline of pluvial Lake Thompson (now Rogers and Rosamond dry lakes). This sequence is typical of exposures along the former lake margins and shows a thin veneer of upper Palmdale alluvium overlying an apparent regressive sequence of lacustrine sediments. Alluvium at the base of the cut may also be upper Palmdale in age, as indicated by the lack of a buried soil at the contact.

region consist of two units of probable late Holocene age. These deposits are less than 5 m thick and occur only near active stream channels. Post-Palmdale I deposits underlie recent floodplains. They exhibit bar-and-swale topography, and their surfaces parallel the modern channel profiles. Soils on these deposits are weakly developed; horizonation (A and Cox) is barely present, and clasts are quite fresh. The younger, post-Palmdale II deposits are inset into post-Palmdale I deposits. They are unvegetated and form the active channels. No soil development is apparent on this unit.

Evidence for Synchronous Deposition of Correlative Units

The sediments of the Antelope Valley can be classified according to stratigraphic and geomorphic criteria, but in order to understand the history of late Quaternary alluvial deposition in the basin, it is important to know if the deposits of any given unit are the same age throughout the region. To date, no material suitable for ^{14}C dating has been found, and other numerical dating techniques have not been applied. However, I believe a case can be made in support of regionally time-correlative units on the basis of stratigraphic evidence and examination of geologic and geophysical properties of the sediments. Hence, all of the deposits within a given major unit probably resulted from a single, widespread aggradational episode that simultaneously af-

fected all or most of the drainages that feed into the Antelope Valley.

Maps of the Quaternary geology of the Antelope Valley (Ponti and Burke, 1980; Ponti and others, 1981) show that stream-terrace deposits of the upper Palmdale unit can be continuously traced from the foothill valleys of both mountain ranges onto the basin floor, where upper Palmdale sediments form broad alluvial fans that grade laterally into one another. Nowhere is there evidence for cross-cutting of the individual fans. Furthermore, multiple surfaces within the upper Palmdale sediments, which would be indicative of subsequent episodes of downcutting and backfilling, are not evident except in areas directly adjacent to active faults and folds. This evidence suggests that many drainages heading in the Transverse Ranges and Tehachapi Mountains were aggrading nearly simultaneously during the latest Pleistocene or early Holocene. Because the upper Palmdale sediments that originated in the Transverse Ranges and the Tehachapi Mountains can be traced continuously between the two ranges, it appears that the deposition of the upper Palmdale unit was not originally generated by uplift along a mountain front. Also, the sequence of upper Quaternary deposits is identical along both mountain ranges. If uplift along mountain fronts was the principal cause of alluvial deposition in this region, alluvial sequences along the two mountain fronts should differ markedly because of their different tectonic settings.

The lower Tylerhorse through lower Palmdale deposits are not as well preserved as the upper Palmdale deposits and therefore cannot be physically traced over large distances. Although it may be assumed by analogy with the upper Palmdale deposits that the older units are also coeval, this cannot be positively established stratigraphically. Establishing the time-correlative nature of the older deposits requires the use of relative dating techniques; this involves the quantification of geologic, geomorphic, and geophysical properties of the sediments that are assumed to be age-dependent. If the variation in any of these age-dependent properties within a given unit is much smaller than the variation in properties among the different units, it is reasonable to conclude that the deposits within each unit are roughly time-correlative (Birkeland and others, 1979; Burke and Birkeland, 1979). The properties chosen for examination include: (1) several quantitative soil development indices, (2) in situ shear-wave velocities, and (3) a quantitative assessment of fan-surface dissection.

Soil Development Indices

Several soil indices, which are quantitative assessments of the degree of soil profile development, were computed for twenty soil profiles selected at locations throughout the basin (Fig. 1, Table 1); thirteen of the soil profiles were described by the Soil Conservation Service (Woodruff and others, 1970), and the rest were developed for this report. In addition, the soils described for this report and some horizons from soils described by Woodruff and others (1970) were sampled for particle size analysis. From the descriptions and laboratory data, eight quantitative indices of soil development were computed for each profile (Table 1). These are: (1) the Harden profile-development index (Harden, 1982); (2) a thickness-normalized profile index taken from Harden and Taylor (1983); (3) the thickness of the argillic horizon; (4) the maximum percent clay increase from the C to the B horizon; (5) the maximum reddening (change in hue) as compared to the parent-material color; (6) the maximum increase in chroma as compared to the parent-material color; (7) the maximum normalized horizon index (Harden, 1982); and (8) the sum of the maximum normalized soil-property indices (Harden, 1982) for clay films, texture, and rubification (soil reddening).

The profile-development index of Harden (1982) is determined by quantifying the difference between each horizon and its parent material for up to ten different soil properties. The index for each soil property is then normalized by dividing the mean value by the maximum value for each property. Next, all of the normalized property indices for each horizon are summed, and the sum is then divided by the number of properties used to develop the index (in this case there were eight); this number is termed the normalized horizon index. The final profile index is computed by multiplying each horizon index by the horizon thickness and then summing the values of each horizon. Soil properties considered in developing the profile index are clay films, total texture, rubification, structure, dry and moist consis-

tency, melanization (darkening), and pH. The larger the profile index value, the better developed the soil. However, the absolute value of the profile index is highly dependent on the total depth of the described soil. To attempt to adjust for this problem, a thickness-normalized profile index (Harden and Taylor, 1983) is computed by dividing the profile index by the total thickness of the described profile. The thickness of the argillic horizon and maximum percent clay increase from the C to the B horizon are taken directly from the soil descriptions and laboratory data, and the index values of reddening (change in hue) and change in chroma are computed by comparing the maximum Munsell color within the soil profile (usually in the B horizon) to the Munsell color of the parent material (Table 1). The maximum normalized horizon index is the value of the best developed horizon in the profile and is derived from computation of the Harden profile index. The numerical value of the maximum normalized soil-property index is obtained by summing the maximum index values for clay films, texture, and rubification within each profile (Harden, 1982).

All of the above indices are presumed to increase in value with increasing soil development, and therefore with time, although the rates at which the indices change through time may vary significantly because of the effects of other soil-forming factors. Plots of each of the soil-index values by stratigraphic unit (Fig. 4) show that each of the indices roughly increases in value as the age of the deposit increases. It is apparent, however, that some of the indices show this trend more clearly than others. For any given index, the values for a given stratigraphic unit commonly overlap with the values for some of the other units. Thus, no single index is capable of completely discriminating among certain of the stratigraphic units.

To determine whether or not a combination of indices could be used to differentiate between the stratigraphic units, a discriminant analysis was performed using each of the soil indices in Table 1 as discriminating variables and the stratigraphic units as groups (Nie and others, 1975). The results of this analysis (Fig. 5) show that the soil indices provide excellent discrimination among the units (canonical correlation for the two discriminant functions = 0.96); the variation among the group centroids is quite large and indicates that the combined soil-index values of the different stratigraphic units are substantially different. The development indices most useful in defining the stratigraphic units are, in order of decreasing importance: (1) thickness of the argillic horizon; (2) maximum reddening as compared to the parent-material color; (3) thickness-normalized profile index; and (4) sum of the maximum normalized soil-property indices for clay films, texture, and rubification. Two indices, the maximum increase in chroma as compared to the parent-material color and the maximum normalized horizon index, added insignificantly to the analysis and were not included as variables in either of the two discriminant functions.

Perhaps the most important result of this analysis is that within the older units (lower Palmdale and Tylerhorse sediments) the discriminant scores for each profile cluster tightly around their

TABLE 1. SOIL-DEVELOPMENT INDICES FOR SOIL-SAMPLE SITES IN THE ANTELOPE VALLEY REGION
[Abbreviations used in column headings: Max., maximum; %, percent; p.m., parent material.]

Site[1]	Location	Unit	Harden profile-development index[2]	Thickness-normalized profile index[3]	Thickness of argillic horizon (cm)	Max. % clay increase from C to B horizon	Max. reddening as compared to p.m. color[4]	Max. chroma increase as compared to p.m. color[5]	Max. normalized horizon index[2]	Sum of max. normalized soil property indices: clay films, texture and rubification[2]
S7-1	1.2 miles NW of Longview Rd. and Ave. W; SW1/4, NE1/4, Sec. 25, T5N, R10W	Post-Palmdale I	11.07	0.0723	0	0	0	1	0.086	0.216
S7-2	0.15 miles N of 130th St. E and Palmdale Blvd.; SE1/4, NE1/4, Sec. 26, T6N, R10W	Post-Palmdale I	10.07	0.0663	0	0	0	1	0.069	0.216
S7-3	0.9 miles S of Valyermo and Pallett Crk. Rds.; NE1/4, NE1/4, Sec. 7, T4N, R9W	Post-Palmdale I	6.57	0.0722	0	0	0	2.5	0.073	0.295
D6-1	On 170th St. W; SW1/4, NW1/4, Sec. 25, T8N, R15W	Upper Palmdale	6.27	0.1028	15	3	0	0	0.119	0.158
D6-2	NE corner, Sec. 9, T7N, R13W	Upper Palmdale	11.35	0.1247	0	0	0	1	0.162	0.438
D6-3	East side freeway cut, Hwy. 14 and Ave. M; SW1/4 SE1/4, Sec. 33, T7N, R12W	Upper Palmdale	6.57	0.0722	0	0	0	2.5	0.073	0.295
S6-1	0.3 mile S of Lancaster Rd. and 205th St. W, SW1/4, SE1/4, Sec. 29, T8N, R15W	Upper Palmdale	16.36	0.0930	0	3	0	2	0.101	0.222
S6-2	Near center of NE1/4, SE1/4, Sec. 8, T5N, R10W	Upper Palmdale	21.64	0.1109	0	0	0	1	0.177	0.327
S6-3	1.6 miles west of Old Liebre Ranch; SW1/4, SW1/4, Sec. 20, T8N, R17W	Upper Palmdale	25.55	0.1265	0	3	0	0	0.135	0.333
S6-4	In SW1/4, SW1/4, Sec. 33, T8N, R10W	Upper Palmdale	17.92	0.1156	0	0	0	1	0.148	0.105
S6-5	560 ft. west of NE corner of Sec. 11, T8N, R17W	Upper Palmdale	24.78	0.1619	0	0	0	0	0.225	0.661
S4-1	At Ave. M-8, and 37th St. W. NE1/4, SE1/4, Sec. 6, T6N, R12W	Lower Palmdale	38.26	0.1747	99	8	2	2	0.282	1.260
S4-2	Near center of NE1/4, NW1/4, Sec. 30, T7N, R13W	Lower Palmdale	31.25	0.1532	79	3	0	0	0.191	0.910
S4-3	0.4 mile west of SW corner of Sec. 26, T9N, R16W	Lower Palmdale	42.92	0.2104	69	5	1	2	0.252	1.554
D3-1[6]	Roadcut on 10th St. W, 1/4 mile N of Ave. Q; SW1/4, SW1/4, Sec. 22 T6N, R12W	Upper Tylerhorse	57.99	0.2660	66	17	2	3.5	0.375	1.716
D3-2	Roadcut on 110th St. W, 1/4 mile S of Ave. K; SW1/4, NW1/4, Sec. 25, T7N, R14W	Upper Tylerhorse	49.78	0.2540	58	8	1	3	0.333	1.353
S3-1	0.3 miles W of 20th St. W, on Ave. O-8; NW1/4, SE 1/4, Sec. 14, T6N, R12W	Upper Tylerhorse	37.17	0.2368	61	18	2	2	0.463	1.714
D2-1	300 ft S of Site D3-2	Middle Tylerhorse	64.23	0.2805	116	16	1	3	0.399	1.584
S2-1	0.1 mile E of Johnson Rd. and 110th St. W; SW1/4 SW1/4, Sec. 25, T7N, R14W	Middle Tylerhorse	48.10	0.2100	112	18	2	1	0.323	1.507
D1-1	50 ft S of dirt road; NE1/4 NW1/4, Sec. 7, T4N, R8W	Lower Tylerhorse	30.61	0.5018	>60	20	2	4	0.563	2.252

[1] "S" sites were described and sampled by the Soil Conservation Service (Woodruff and others, 1970). The percent clay increase for these sites is estimated from descriptions and laboratory data from Woodruff and others (1970). "D" sites were sampled and described by the author.

[2] Computed from Harden (1982).

[3] Designated as the profile index (Harden, 1982) divided by depth of soil profile.

[4] Computed from the Munsell dry color designation, with each reddening of hue corresponding to an index value of 1. For example, the index value for a reddening from 10YR to 7.5YR would equal 1.

[5] Computed from the Munsell dry color designation, with each increase in chroma corresponding to an index value of 1. For example, the index value for an increase in chroma from /4 to /6 would equal 2.

[6] Contains abundant schist clasts.

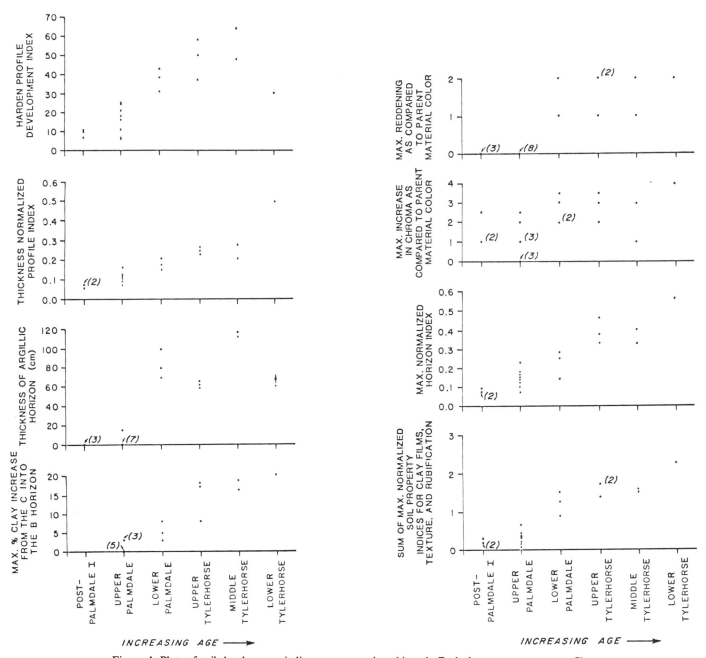

Figure 4. Plots of soil development indices versus stratigraphic unit. Each dot represents one profile.
Numbers in parentheses indicate the number of profiles that share the same or similar values.

respective group centroids. In fact, the clustering appears to be even tighter in the older units than in the upper Palmdale and post-Palmdale sediments. This suggests that the variability of the soils formed on the lower Palmdale and Tylerhorse sediments is even less than the variability of soils formed on upper Palmdale deposits. Because we can demonstrate stratigraphically that the soils formed on the upper Palmdale deposits are about the same age, the above evidence indicates that the older deposits are probably time-correlative as well.

The significance of the time discriminant analysis is suspect due to the paucity of data from older sediments. However, from the data presently available, it is apparent that the degree of development of soils on a given unit is considerably less variable (in spite of some differences in rainfall and parent material over the region) than the degree of development among soils on different units. If deposits of the same unit were not time-correlative across the basin, there should be even more variation in the soils.

In Situ Shear-Wave Velocities

Although many of the physical factors that control shear-wave velocity are not completely understood, it appears that the

D. J. Ponti

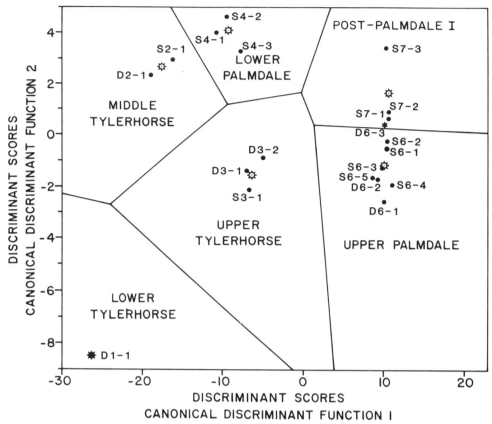

Figure 5. Plot of discriminant scores computed for 20 soil sites within the Antelope Valley. The variables entered into the discriminant analysis are those listed in Table 1. Canonical discriminant function 1 discriminates between the groups principally on the basis of thickness of the argillic horizon, and on soil color (hue). Canonical discriminant function 2 discriminates on the weighted mean profile index and the sum of maximum clay films, texture, and rubification values. Open stars are the group centroids. The lines dividing up the plot mark regions such that if a site falls on a line it has equal probability of belonging to the unit on either side of the line. One site, D6-3 (marked by the asterisk), is misclassified; this deposit is considered to be upper Palmdale, but has a soil with properties more like those of soils on post-Palmdale deposits.

in situ shear-wave velocity of unconsolidated to poorly consolidated sediment is primarily a function of cementation, void ratio, and sediment texture (Fumal, 1978). Because the degree of cementation and the void ratio may be age-dependent, a study was conducted in the Antelope Valley to see if there is a relation between shear-wave velocities and the age of deposits (Fumal and others, 1982). Nine holes were drilled in the basin to depths of 30 m, and the results (Fig. 6) show a distinct clustering of in situ shear-wave velocities within each of the major stratigraphic units sampled.

Studies in the San Francisco and Los Angeles regions (Fumal, 1978; Fumal and others, 1982) show that shear-wave velocity progressively increases in successively older deposits. Older sediments tend to have lower void ratios and are more highly cemented; this tends to improve grain-to-grain contacts and hence increases shear-wave velocity. These studies also show, however, that within deposits of a single age the velocities can vary significantly with texture and depth of burial. Clayey sediments generally have lower velocities than do gravels, and sands

have intermediate velocities. The most variation can be seen in the clays, where velocities appear to be considerably affected by overconsolidation (Fumal, 1978). However, within the range of textures from gravelly sand to sandy loam (characteristic of most Antelope Valley deposits), seismic velocities within a deposit appear to be fairly consistent (Fumal, 1978, and pers. commun., 1982). Also, within the upper 30 m, velocities are not substantially affected by depth of burial (T. Fumal, pers. commun., 1982).

Because the texture and depth of burial of the sandy alluvial sediments in the Antelope Valley probably do not significantly affect the measured in situ shear-wave velocities, it is likely that the substantially different seismic velocities in the units are due to differences in age. Tight clustering of seismic velocities within the upper and middle Tylerhorse sediments suggests near-synchronous deposition. Some overlap in velocities occurs in sediments of the upper and lower Palmdale units. This overlap may be due to random variability; however, the high velocity values for two of the upper Palmdale sites close to the San Andreas fault may be

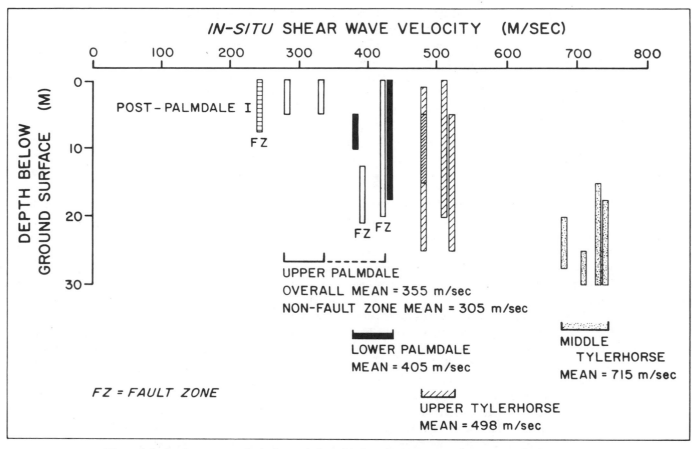

Figure 6. In situ shear-wave velocity intervals from 9 holes drilled into upper Quaternary alluvium in the Antelope Valley region. Velocities cluster within each unit with the exception of two upper Palmdale sites. Holes drilled within several hundred meters of the active trace of the San Andreas fault are designated by the symbol FZ. Data from Fumal and others (1982).

related to overconsolidation caused by high ground-water conditions and high seismic accelerations in and near the fault zone.

Fan and Terrace Surface Dissection

After the deposition of an alluvial fan or terrace, subsequent relative uplift and stream downcutting will cause the original surface of the deposit to be eroded. The amount of erosion or dissection of the fan or terrace surface is due, in part, to climate or tectonics or a combination of both, but is ultimately a function of age.

The effects of time on the dissection of the Antelope Valley deposits were assessed for seventeen terraces selected at random within the basin. For each of these terraces, the degree of dissection was quantified by means of a dissection index (DI). This number is the ratio of the length of a contour line across a terrace surface to the straight-line distance across the terrace, and is a direct measure of the amount of gullying that has occurred on what was originally presumed to have been a smooth surface. For each terrace, five contour lines were measured from a topographic map (contour interval equals 10 ft or 3 m), and the DIs were averaged to arrive at a value for that terrace. The results (Fig. 7)

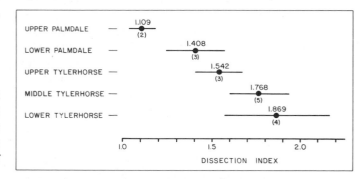

Figure 7. Average dissection index values (ratio of length of a contour line across a terrace surface to the straight-line distance across the terrace) for fans and terraces in the Antelope Valley region. Error bars are ± one standard deviation. Numbers in parentheses are the number of surfaces examined in each analysis.

reveal that the mean DIs increase with age of the deposit, as expected, but the large standard deviations indicate that there is considerable variation within each unit. These variations in the DIs are likely caused by local differences in tectonic uplift rates, texture, or drainage basin evolution and do not necessarily preclude the usefulness of such an index for correlation purposes.

Although not conclusive, combined evidence from stratigraphic data, soil indices, shear-wave velocity, and degree of terrace dissection strongly suggests that deposits mapped as a single unit from geologic and geomorphic criteria are essentially time-correlative throughout the Antelope Valley. This presumption is based on limited data that suggest the variability of certain soil and engineering properties is substantially greater among units that within units.

Controls on Deposition

As stated previously, the existence of a consistent sequence of time-correlative deposits over a large area such as the Antelope Valley suggests that major episodes of alluviation are probably controlled by changes in climate. In support of this theory, other evidence indicates that deposition of the principal alluvial units occurred during short periods of time, and sedimentological evidence points to a shift to different hydrologic conditions during principal periods of alluviation.

Buried soils, which are widespread in many of the Quaternary deposits in the Antelope Valley, suggest that significant hiatuses occurred between deposition of the alluvial units. In many places, profile development of the buried soils superficially appears quite comparable to that of surface soils on deposits of equivalent age (Fig. 8). Buried and surface soils at sites D3-2 and D2-1, which are located on Johnson Road about 18 km west of Palmdale (Fig. 1), were sampled for laboratory analysis; although these analyses show that many of the gross morphologies are similar, there are some significant differences between the buried and surface occurrences of these soils. At site D2-1, a soil is formed on middle Tylerhorse sediment. About 100 m away, at site D3-2, the same soil occurs beneath sediments of upper Tylerhorse age. Soil data from these sites (Fig. 9) reveal that the buried argillic horizon at site D3-2 is similar to the argillic horizon at site D2-1 with respect to thickness and color, but the buried horizons contain only about half the percentage of clay that the horizon at site D2-1 contains and have amounts of clay similar to that of the surface argillic horizon at site D3-2. Values for pH and bulk density also are different between the equivalent buried and surface horizons and reflect conditions of reduced leaching and less intense pedogenesis in the buried soil. The soil at site D2-1 is estimated to be about 250,000 years old, by correlation with deposits in the San Joaquin Valley of central California (Marchand and Allwardt, 1981). The surface soil at site D3-2 is estimated in the same manner to be about 140,000 years old; hence, approximately 110,000 years passed before the lower soil at site D3-2 was buried by upper Tylerhorse deposits. The development of the buried argillic horizon is quite similar to surface argillic horizons that have formed on upper Tylerhorse deposits and probably represents at least 100,000 years of development; therefore, if most pedogenesis of the middle Tylerhorse deposit at site D3-2 had ceased after burial, deposition of the upper Tylerhorse unit would not have required a great deal of time. Buried soil relationships such as these have also been noted in other regions

(Janda and Croft, 1967; Tinsley and others, 1982) and have been interpreted to be indicative of rapid deposition.

Particle-size data from deposits in the region suggest that the character of the Palmdale and Tylerhorse sediments differs from the post-Palmdale sediments that were deposited in the late Holocene. Particle-size analyses from 144 sites selected at random in upper Palmdale and post-Palmdale I deposits show that, on average, significantly coarser mean particle sizes and significantly higher standard deviations (poorer sorting) occur in the post-Palmdale sediments (Fig. 10). Post-Palmdale I sediments are also slightly more skewed toward the finer size fractions than are the upper Palmdale sediments.

Similar results are derived from the sand-fraction data of five of the soil sites plotted in Figure 1. The distribution of particle sizes within the sand fraction in the soil is considerably different from the sand-fraction distribution in sediments from the channel and active floodplain adjacent to the soil site (Fig. 11). Like the bulk-sediment analysis, the sand distributions for the active channels and floodplains (post-Palmdale I and II sediments) are coarser grained and more skewed toward the fine fraction than is the sand distribution in the adjacent soil sample. Because sands are probably not altered significantly during pedogenesis (Birkeland, 1974), the differences between the soil and floodplain samples are likely to be indicative of different parent materials.

These differences in parent material probably reflect a shift in hydrologic conditions during deposition of the major alluvial units. Modern deposition occurs locally along channels and floodplains primarily during storm-generated flash-flood events; the particle-size distribution of the modern sediments (post-Palmdale I and II) is consistent with this type of deposition. In contrast, the finer grained and better sorted Palmdale and Tylerhorse deposits probably indicate a much more stable hydrologic regime, characterized by higher base flows and somewhat lower peak discharges. Based on present evidence, the periods of time characterized by large-scale basin-wide deposition were probably wetter than present (in order to support the higher stream flows) and probably quite short-lived.

To summarize, in the Antelope Valley, as has also been inferred for the San Joaquin Valley (Janda and Croft, 1967), most of the late Quaternary appears to have consisted of periods of landscape stability and soil formation, punctuated by limited deposition along channels and floodplains during flash-flood events and erosion in tectonically active areas. Times of major alluviation appear to have occurred only during brief episodes, probably as a result of climatic aberrations that produced higher stream base flows than occur at present.

AGE ESTIMATES FOR THE ALLUVIAL SEQUENCE OF THE ANTELOPE VALLEY

Present data suggest that the principal alluvial deposits of the Antelope Valley range in age from Pleistocene to early Holocene. The Harold Formation, which unconformably(?) underlies the

Figure 8. Stream cut on the south flank of the Tehachapi Mountains (Tylerhorse Canyon quadrangle; NE ¼, SE ¼, Sec. 31, T10N, R14W) showing upper Palmdale sediments overlying sediments from the upper Tylerhorse unit. Gray zone near the figure is the argillic horizon formed on the upper Tylerhorse sediment prior to burial. The thickness (about 50 cm) and color (7.5YR 5/4) of this horizon are comparable to those of surface soils on this unit.

Tylerhorse and Palmdale sediments, is approximately 700,000 years old to 1.2 m.y. old based on a vertebrate fossil assemblage collected from a section of the Harold Formation that is adjacent to the Nadeau fault near Palmdale (C. Repenning, written commun., 1982). Therefore, an age of around 700,000 years to 1.2 m.y. would be the oldest possible date for the lower Tylerhorse deposits. Deposits of the upper Palmdale unit appear to interfinger with, and overlie, lacustrine sediments from pluvial Lake Thompson (Rogers and Rosamond dry lakes). Age estimates for the last pluvial highstand throughout the Mojave Desert region and the western U.S. range from 12,000 to 25,000 years ago, with most estimates between 12,000 and 14,000 years ago (Smith, 1968; Benson, 1978; Lajoie, 1982; and Scott and others, 1983). Ostracods from a deep-water clay in Koehn Lake, located only 50 km north of Rogers Lake, yield a ^{14}C date of 14,700 ± 130 years (Burke, 1979); two ^{14}C dates on lithoid tufa from an

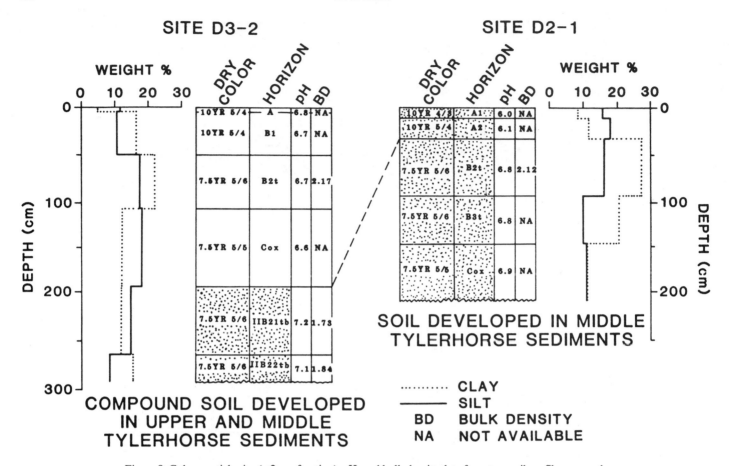

Figure 9. Color, particle size (<2mm fraction), pH, and bulk density data from two soil profiles exposed along Johnson Road west of Palmdale, on the north flank of the Transverse Ranges. The buried argillic horizon at site D3-2 formed in middle Tylerhorse sediments and corresponds to the surface soil formed at site D2-1. The pH values were obtained using a field (Trug) pH kit; bulk density values were obtained using the soil aggregate method.

offshore bar apparently deposited during the last highstand of Koehn Lake give ages of 12,700 ± 100 years and 13,460 ± 80 years (Clark and Lajoie, 1974). These dates suggest that upper Palmdale sediments range in age from latest Pleistocene to early Holocene.

The only way to estimate the ages of the other units is to correlate them with dated deposits elsewhere. Because principal alluvial deposition in the Antelope Valley appears to have been controlled by regional climate change, the depositional sequence of these deposits should be similar to that found in other areas. A sequence of dated alluvial deposits in the eastern San Joaquin Valley, located about 300 km to the north, appears similar to the sequence in the Antelope Valley. The two major units of the San Joaquin Valley, the Riverbank and Modesto Formations of Pleistocene age, were deposited mainly as a result of glacial outwash from the Sierra Nevada, thereby establishing climatic change as a major influence on their deposition. Like the Antelope Valley deposits, the Riverbank and Modesto sediments are predominantly of granitic origin and have been subjected to weathering conditions under an arid to semi-arid climate. The only principal

difference between the two regions is that parent-material textures of the deposits are finer-grained in the San Joaquin Valley, ranging from sandy loam to silt loam.

A comparison of the two depositional sequences reveals some striking similarities (Table 2). First of all, if one assumes that the middle Palmdale unit does not represent a separate episode of aggradation, there is a remarkable one-to-one correspondence of units. Secondly, the age range for the upper member of the Modesto corresponds very closely to what is inferred to be the age of the upper Palmdale sediments. Also, the limiting maximum age of 700,000 years to 1.2 m.y. for the lower Tylerhorse deposits is not at all incompatible with the 600,000 year age estimate for the upper unit of the Turlock Lake Formation (Marchand and Allwardt, 1981). Finally, many of the same soil series are mapped by the Soil Conservation Service on corresponding deposits, and the geomorphic characteristics of the terrace surfaces are also fairly similar (Ponti, 1980).

The mere existence of very similar alluvial sequences in these two areas does not, however, prove that the deposits are time-correlative. The Antelope Valley sediments are not glacial

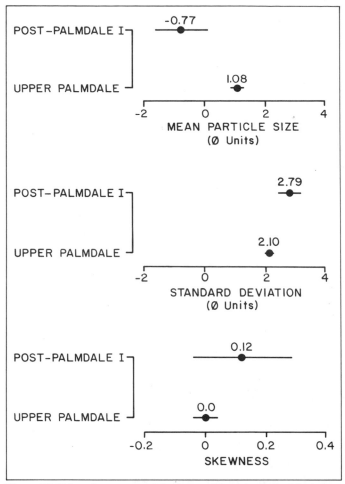

Figure 10. Values for mean particle size, particle-size standard deviation, and skewness for 144 sites selected at random in upper Palmdale and post-Palmdale sediments. Statistics are computed from Folk (1974), data are from Ponti (1980).

outwash deposits, and, therefore, while their deposition may be climatically controlled, it does not necessarily follow that the two sequences have to correspond in time. However, on the basis of stratigraphic evidence from the San Joaquin Valley and the similarity among soil development indices for soils on deposits of the two regions, these two sequences probably correspond fairly closely in time, perhaps within a few thousand years.

Within the eastern San Joaquin Valley, alluvium from non-glaciated drainage basins apparently slightly post-dates glacial outwash (Ponti and others, 1980). This relationship can be demonstrated for deposits of late Modesto, early Modesto, and late Riverbank age, where non-glacial alluvium overlies glacial outwash with no evidence of a buried soil. This relationship is also observed by Lettis (1982), who demonstrated that non-glacial alluvium derived from the western San Joaquin Valley post-dates Sierra Nevada derived glacial outwash without evidence of a buried soil. These relationships indicate that an episode of non-glacial alluviation occurred during several glacial to interglacial transitions.

Other evidence in support of the approximate contemporaneity of the two sequences is that for the soil development indices that are most useful for distinguishing deposits of different ages the index values for corresponding units are quite similar (Table 3). The one major difference is that index values for thickness of the argillic horizon are substantially different between the two regions. This difference is probably due to the much higher clay content in the parent material of the San Joaquin Valley soils.

If the correlation between the Antelope Valley and the eastern San Joaquin Valley deposits is sound, the Tylerhorse and Palmdale alluvial sediments are probably equivalent to, or slightly younger than, the Riverbank and Modesto Formations, respectively.

TABLE 2. COMPARISON OF ANTELOPE VALLEY AND EASTERN SAN JOAQUIN VALLEY ALLUVIAL SEQUENCES

Eastern San Joaquin Valley Unit[1]	Characteristic Soil Series (Arkosic, no hardpan)[1]	Age Estimate (yrs)[1]	Dating Method[1]	Presumed Equivalent Antelope Valley Unit[2]	Characteristic Soil Series (Arkosic, no hardpan)[2]	Age Estimate (years)[2,5]	Dating Method[2]
Post-Modesto IV alluvium	Riverwash	<100	Active channels	Post-Palmdale II	Riverwash	<100	Active channels
Post-Modesto II alluvium	Tujunga, Grangeville	3,000	[14]C	Post-Palmdale I	Arizo, Soboba	Late Holocene	Stratigraphic, limitations, soil development
Modesto Fm., upper member	Hanford, Cajon	9,000-14,000	[14]C, stratigraphic limitations	Upper Palmdale sediments	Hanford	<12,000-25,000	Stratigraphic limitations
Modesto Fm., lower member	Greenfield, Borden	30,000-90,000	Limiting ^{14}C dates, uranium trend	Lower Palmdale sediments	Greenfield	Pleistocene	Soil development
Riverbank Fm., upper member	Snelling, Ramona weak variant	About 140,000	Uranium trend	Upper Tylerhorse sediments	Ramona, Mohave	>100,000	Soil development
Riverbank Fm., middle member	Snelling, Ramona normal variant	About 250,000	Uranium trend, fossils	Middle Tylerhorse sediments	Ramona	>200,000	Soil development, buried soils
Riverbank Fm., lower member	Snelling, Ramona strong variant	330,000-480,000	Stratigraphic limitations and correlation with marine oxygen-isotope record[3]	Lower Tylerhorse sediments	Ramona, eroded	>300,000	Soil development, buried soils
Turlock Lake Fm., upper unit	Montpelier	About 600,000	K-Ar on Friant Ash[1]	Harold Formation	--------------	700,000-1,200,000	Fossils[4]

[1] Marchand and Allwardt (1981).
[2] Ponti (1980) and this report.
[3] Shackleton and Opdyke (1973).
[4] Repenning (written commun., 1982).
[5] These age estimates are based solely on stratigraphic and soil development data and are not dependent on correlation with any other area.

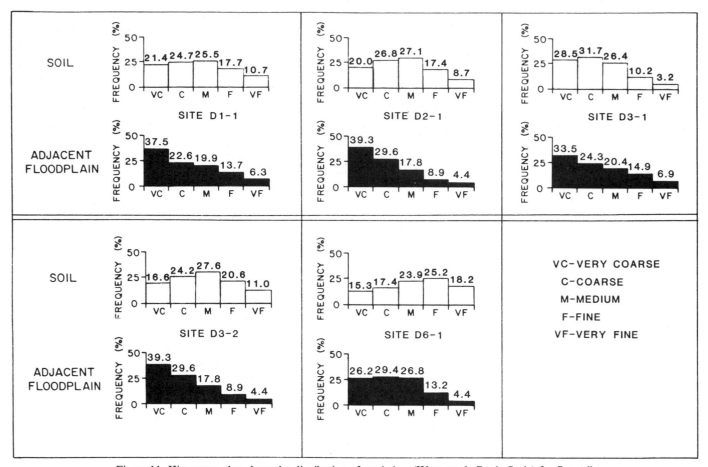

Figure 11. Histograms that show the distribution of sand sizes (Wentworth Grade Scale) for five soil sites in the Antelope Valley. Upper histograms (light gray) show the average sand distributions within the soil profile. Lower histograms (dark gray) show the distribution of sand sizes from a bulk sample taken from sediments within the active floodplain or channel (post-Palmdale I or II) directly adjacent to the soil site.

TABLE 3. COMPARISON OF AVERAGE SOIL INDICES FOR CORRESPONDING STRATIGRAPHIC UNITS IN THE ANTELOPE VALLEY AND MERCED RIVER (EASTERN SAN JOAQUIN VALLEY) AREAS

		Merced River[1]	Antelope Valley	Merced River[1]	Antelope Valley	Merced River[1]	Antelope Valley	Merced River[1]	Antelope Valley	Merced River[1]	Antelope Valley
Stratigraphic unit		Upper member Modesto	Upper Palmdale	Lower member Modesto	Lower Palmdale	Upper member Riverbank	Upper Tylerhorse	Middle member Riverbank	Middle Tylerhorse	Lower member Riverbank	Lower Tylerhorse
Number of profiles		2	8	1	3	2	3	2	2	1[2]	1[3]
Thickness- normalized profile index	Mean	0.107	0.114	0.162	0.179	0.291	0.252	0.271	0.245	0.299	0.502
	Std. Dev.	0.040	0.026	---	0.029	0.002	0.029	0.033	0.147	---	---
Thickness of argillic horizon (cm)	Mean	0	0	89	82.3	106.5	61.7	233.0	114.0	119.0	>60
	Std. Dev.	0	0	---	15.3	33.2	4.0	117.4	2.8	---	---
Maximum reddening as compared to parent material color	Mean	0	0	0	1	1	1.7	1	1.5	2	2
	Std. Dev.	0	0	---	1	---	0.6	0	0.7	---	---
Sum of maximum normalized soil property indices for clay films, texture, and rubification	Mean	0.33	0.32	1.52	1.24	1.53	1.59	1.63	1.55	1.76	2.25
	Std. Dev.	0.24	0.18	---	0.32	0.06	0.32	0.18	0.21	---	---

[1] Data from Harden (1982).
[2] This soil is formed in sediments only 220 cm thick that overlie the Turlock Lake Formation. Weathering probably proceeded into the older sediments, but is masked by a buried soil horizon.
[3] Profile description ends at 61 cm.

SUMMARY AND CONCLUSIONS

Upper Quaternary sediments of the Antelope Valley consist of six principal deposits representing five or possibly six episodes of major fan and terrace deposition. These deposits comprise the Tylerhorse and Palmdale sediments. Tylerhorse sediments are mainly characterized by well-developed reddish brown soils and are extensively dissected. Palmdale sediments have weakly to moderately developed soils and form broad, relatively undissected terraces and alluvial fans. Two other units of post-Palmdale age are mapped in the valley, but they are thin and of limited extent.

Upper Palmdale deposits are continuous from the foothill valleys of both the Tehachapi Mountains and Transverse Ranges out onto the basin floor, and they appear to have been deposited throughout the region at roughly the same time. The presence of a widespread deposit of a single age indicates that all drainages along the Transverse Ranges and Tehachapi mountain fronts were aggrading at the same time, and this probably occurred as a result of regional climate change. Similarity of soils, shear-wave velocities, and degree of fan surface dissection among older deposits suggest that correlation is possible and that alluviation was synchronous across the basin during deposition of the lower Palmdale and Tylerhorse units.

Evidence from buried soils suggests that the periods of major aggradation were short. Particle-size data indicate that the hydrologic conditions during principal alluviation were different from present conditions; stream base flows were probably higher while peak flows were probably lower during deposition of the major units.

The Palmdale and Tylerhorse sediments appear to correlate with the Riverbank and Modesto Formations of the eastern San Joaquin Valley. Based on this correlation and on stratigraphic and soil development constraints, the ages of the Palmdale and Tylerhorse sediments are as follows: (1) upper Palmdale, 8,000 to 14,000 years; (2) lower Palmdale, 40,000 to 70,000 years(?); (3) upper Tylerhorse, about 140,000 years; (4) middle Tylerhorse, about 250,000 years; and (5) lower Tylerhorse, about 330,000 to 480,000 years.

These ages correspond roughly with glacial stages from the deep-sea oxygen-isotope record (Shackleton and Opdyke, 1973). Because of relationships in the San Joaquin Valley that suggest that deposition of non-glacial alluvium slightly post-dates glacial outwash, it appears likely that the non-glacial Tylerhorse and Palmdale sediments were deposited during transitions from the glacial to interglacial periods reflected in the marine oxygen-isotope records. Hence, it is possible that major alluviation in the Antelope Valley, and in perhaps much of the southwestern U.S., has resulted from hydrologic change associated with worldwide climatic variations.

REFERENCES CITED

Benson, L. V., 1978, Fluctuation in the level of pluvial Lake Lahontan during the last 40,000 years: Quaternary Research, v. 9, p. 300–318.

Birkeland, P. W., 1974, Pedology, weathering, and geomorphological research: New York, Oxford University Press, 285 p.

Birkeland, P. W., Colman, S. M., Burke, R. M., Shroba, R. R., and Meierding, T. C., 1979, Nomenclature of alpine glacial deposits, or, what's in a name?: Geology, v. 7, p. 532–536.

Burke, D. B., 1979, Log of a trench in the Garlock fault zone, Fremont Valley, California: U.S. Geological Survey Miscellaneous Field Studies Map MF-1028, scale 1:20.

Burke, R. M., and Birkeland, P. W., 1979, Reevaluation of multiparameter relative dating techniques and their application to the glacial sequence along the eastern escarpment of the Sierra Nevada, California: Quaternary Research, v. 11, p. 21–51.

Clark, M. M., and Lajoie, K. R., 1974, Holocene behavior of the Garlock fault [abs.]: Geological Society of America Abstracts with Programs, v. 6, p. 156–157.

Dibblee, T. W., Jr., 1967, Areal geology of the western Mojave Desert, California: U.S. Geological Survey Professional Paper 522, 153 p.

Folk, R. L., 1974, Petrology of sedimentary rocks: Austin, Hemphill Publishing Company, 182 p.

Fumal, T. E., 1978, Correlations between seismic wave velocities and physical properties of geologic materials, San Francisco Bay region, California: U.S. Geological Survey Open-File Report 78-1067, 114 p.

Fumal, T. E., Gibbs, J. F., and Roth, E. F., 1982, In situ measurements of seismic velocity at 22 locations in the Los Angeles, California, region: U.S. Geological Survey Open-File Report 82-833, 138 p.

Gile, L. H., Hawley, J. W., and Grossman, R. B., 1981, Soils and geomorphology in the Basin and Range area of southern New Mexico—Guidebook to the Desert Project: New Mexico Bureau of Mines and Mineral Resources Memoir 39, 222 p.

Harden, J. W., 1982, A quantitative index of soil development from field descriptions: Examples from a chronosequence in central California: Geoderma, v. 28, no. 1, p. 1–28.

Harden, J. W., and Taylor, E. M., 1983, A quantitative comparison of soil development in four climatic regimes: Quaternary Research, v. 20, p. 342–359.

Janda, R. J., and Croft, M. G., 1967, The stratigraphic significance of a sequence of noncalcic brown soils formed on the Quaternary alluvium of the northeastern San Joaquin Valley, California, in Morrison, R. B., and Wright, H. E., Jr., eds., Quaternary Soils: Proceedings, International Association for Quaternary Research (INQUA), VII Congress, Reno, Nevada, v. 9 (Desert Research Institute), p. 157–190.

Jenny, H., 1941, Factors of soil formation: New York, McGraw Hill Book Company, 281 p.

Lajoie, K. R., 1982, Late Quaternary glacio-lacustrine chronology, Mono Basin, California [abs.]: Geological Society of America Abstracts with Programs, v. 14, p. 179.

Lettis, W. R., 1982, Late Cenozoic stratigraphy and structure of the western margin of the central San Joaquin Valley, California: U.S. Geological Survey Open-File Report 82-526, 203 p.

Marchand, D. E., and Allwardt, A., 1981, Late Cenozoic stratigraphic units, northeastern San Joaquin Valley, California: U.S. Geological Survey Bulletin 1470, 70 p.

Nie, N. H., Hull, C. H., Jenkins, J. G., Steinbrenner, K., and Bent, D. H., 1975, Statistical package for the social sciences, second edition: New York, McGraw Hill Book Company, 567 p.

Noble, L. F., 1953, Geology of the Pearland quadrangle, California: U.S. Geological Survey Geologic Quadrangle Map GQ-24, scale 1:24,000.

——1954, Geology of the Valyermo quadrangle and vicinity, California: U.S. Geological Survey Geologic Quadrangle Map GQ-50, scale 1:24,000.

Ponti, D. J., 1980, Stratigraphy and engineering characteristics of upper Quaternary sediments in the eastern Antelope Valley and vicinity, California [M.S. Thesis]: Stanford, Stanford University, 157 p.

Ponti, D. J., and Burke, D. B., 1980, Map showing Quaternary geology of the eastern Antelope Valley and vicinity, California: U.S. Geological Survey Open-File Report 80-1064, scale 1:62,500.

Ponti, D. J., Burke, D. B., and Hedel, C. W., 1981, Map showing Quaternary geology of the central Antelope Valley and vicinity, California: U.S. Geological Survey Open-File Report 81-737, scale 1:62,500.

Ponti, D. J., Burke, D. B., Marchand, D. E., Atwater, B. F., and Helley, E. J., 1980, Evidence for correlation and climatic control of sequences of late Quaternary alluvium in California [abs.]: Geological Society of America Abstracts with Programs, v. 12, p. 501.

Richmond, G. M., 1962, Quaternary stratigraphy of the La Sal Mountains, Utah: U.S. Geological Survey Professional Paper 324, 135 p.

Scott, G. R., 1963, Quaternary geology and geomorphic history of the Kassler quadrangle, Colorado: U.S. Geological Survey Professional Paper 421-A, 70 p.

Scott, W. E., McCoy, W. D., Shroba, R. R., and Rubin, M., 1983, Reinterpretation of the last two cycles of Lake Bonneville, western United States: Quaternary Research, v. 20, p. 261–285.

Shackleton, N. J., and Opdyke, N. D., 1973, Oxygen isotope and paleomagnetic stratigraphy of equatorial Pacific core V28-238: Oxygen isotope temperatures and ice volumes on a 10^5 year and 10^6 year scale: Quaternary Research, v. 3, p. 39–55.

Smith, G. I., 1968, Late Quaternary geologic and climatic history of Searles Lake, southeastern California, in Morrison, R. B., and Wright, H. E., Jr., eds., Means of correlation of Quaternary successions: Proceedings, International Association for Quaternary Research (INQUA), VII Congress, Salt Lake City, Utah, v. 8 (University of Utah Press), p. 293–310.

Tinsley, J. C., Matti, J. C., and McFadden, L. D., eds., 1982, Late Quaternary pedogenesis and alluvial chronologies of the Los Angeles and San Gabriel Mountain areas, southern California, and Holocene faulting and alluvial stratigraphy within the Cucamonga fault zone—a preliminary view: Geological Society of America Cordilleran Section 78th Annual Meeting, Guidebook, Field Trip No. 12, 44 p.

Woodburne, M. O., 1975, Cenozoic stratigraphy of the Transverse Ranges and adjacent areas, southern California: Geological Society of America Special Paper 162, 91 p.

Woodburne, M. O., and Golz, D. J., 1972, Stratigraphy of the Punchbowl Formation, Cajon Valley, southern California: California University Publications in Geological Sciences, v. 92, 73 p.

Woodruff, G. A., McCoy, W. J., and Sheldon, W. B., 1970, Soil survey of the Antelope Valley area, California: U.S. Soil Conservation Service, 187 p., map scale 1:24,000.

MANUSCRIPT ACCEPTED BY THE SOCIETY JANUARY 12, 1985

Geological Society of America
Special Paper 203
1985

Late Cenozoic stratigraphy and structure of the west margin of the central San Joaquin Valley, California

William R. Lettis
U.S. Geological Survey
345 Middlefield Road
Menlo Park, California 94025

ABSTRACT

Upper Pliocene and Quaternary deposits were mapped in an area of 1,800 km^2 in the west-central San Joaquin Valley and adjacent Diablo Range. The upper Pliocene and Pleistocene Tulare Formation, which consists of alluvial sand, gravel, silt, and clay and locally dips 20°, is overlain by a sequence of six units, each 0 to 20 m thick, each the result of an episode of deposition of alluvium from the Diablo Range. Erosional unconformities and soils between the units record the intervening periods. The units consist of unconsolidated gravel, sand, silt, and clay, with textures and sedimentary structures indicating deposition primarily by flowing water and secondardily by mudflows. The lower three units are grouped into the informally designated alluvium of Los Banos, of middle and late Pleistocene age, the two middle units are grouped into the upper Pleistocene alluvium of San Luis Ranch, and the uppermost unit is the Holocene alluvium of Patterson. Holocene arkosic alluvium, derived from the Sierra Nevada and deposited in flood basins along the San Joaquin River, is informally named the alluvium of Dos Palos.

The two older units of the alluvium of Los Banos are coeval with broad pediment remnants preserved across the foothills, which indicates that the present elevation of the foothills is due to late Quaternary deformation. These surfaces are deformed into a series of broad, gentle northeast-trending folds that have been displaced more than 100 m along three northwest-trending fault systems. The Ortigalita fault displaces Holocene alluvium and has predominantly strike-slip displacement. The O'Neill fault system is a group of small reverse faults whose fault planes coincide with bedding in the northeast-dipping Great Valley sequence; these faults are interpreted to be sympathetic displacements associated with continued uplift and northeastward tilting of the foothills. The San Joaquin fault, at the foothill-valley margin, vertically displaces the pediments as much as 140 m. The orientation of the fault plane and the magnitude of lateral displacement, however, are not known. Neither the San Joaquin fault nor the O'Neill fault appears to displace the alluvium of San Luis Ranch or younger alluvium.

INTRODUCTION

The late Cenozoic tectonic and climatic history of the San Joaquin Valley and adjacent Sierra Nevada and Diablo Range is recorded in a thick sequence of upper Cenozoic alluvial-fan and terrace deposits in the San Joaquin Valley. These deposits have been studied extensively in the northeastern and eastern San Joaquin Valley by Gale and others (1939), Arkley (1962), Davis and Hall (1958), Janda (1965, 1966), Helley (1967), Marchand (1977), and Marchand and Allwardt (1981), and more recently in the west-central San Joaquin Valley by Lettis (1982), where their detailed stratigraphy has proved to be a valuable aid in deciphering the late Cenozoic tectonic and climatic history of the region.

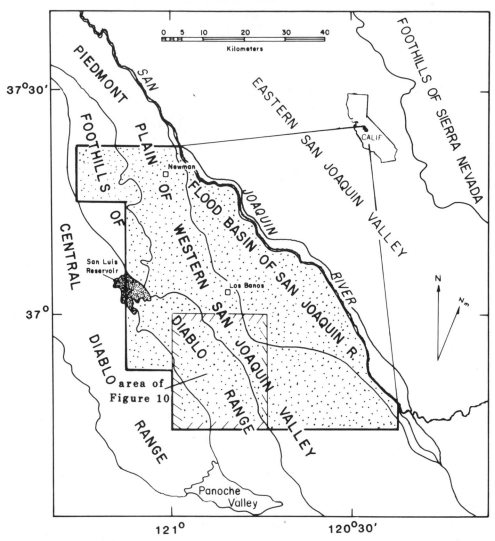

Figure 1. Index map showing location of study area and physiographic divisions of San Joaquin Valley and Diablo Range as recognized in this report.

Alternations of glacial and interglacial conditions in the Sierra Nevada have created a sharply defined chronologic sequence of glacial-outwash alluvial deposits in the eastern San Joaquin Valley (Marchand and Allwardt, 1981). These deposits form a sequence of inset stream terraces and nested alluvial fans near the foothills of the Sierra Nevada and a series of overlapping alluvial fans toward the valley axis in response to continued westward tilting of the Sierran block (Marchand, 1977).

The greater subtlety of the effects of Quaternary climatic changes in the nonglaciated Diablo Range, as opposed to the Sierra Nevada, makes a distinct chronology of alluvial deposits more difficult to establish in the western San Joaquin Valley (Lettis, 1982). This absence of a readily recognizable chronosequence, together with a complex history of Quaternary deformation in the Diablo Range, has generally discouraged detailed stratigraphic study of the upper Cenozoic deposits in the western San Joaquin Valley.

This report briefly describes the upper Cenozoic deposits in

the west-central San Joaquin Valley, uses their lithologic, stratigraphic, and geomorphic relations to interpret the tectonic and climatic history of the western San Joaquin Valley and adjacent Diablo Range, and presents a hypothetical climatic model that I believe governs the deposition of alluvium in the San Joaquin Valley. This study is part of a continuing regional investigation of the late Cenozoic history of the San Joaquin Valley supported by the Reactor Hazards Research Program of the U.S. Geological Survey.

Physical Setting

The west-central San Joaquin Valley and adjacent foothills of the Diablo Range encompass an area of 1,800 km² west of the lower San Joaquin River between lat 36°45′ and 37°22′30″ N., in western Fresno, Merced, and Stanislaus Counties and easternmost San Benito County, California (Fig. 1). The study area straddles the boundary between the Great Valley and Coast

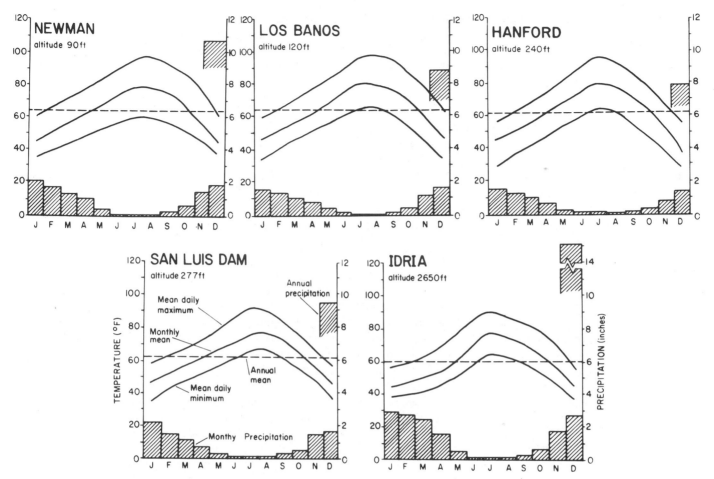

Figure 2. Temperature and precipitation data for five weather stations in San Joaquin Valley and Diablo Range, compiled for the years 1967 through 1976. Data from U.S. National Oceanic and Atmospheric Administration (1967–1976).

Ranges physiographic provinces of California, as described by Jenkins (1943); these provinces are further divided into four natural topographic regions trending northwest through the study area (Fig. 1). The Coast Ranges province is divided into the central Diablo Range and the foothills of the Diablo Range, whereas the Great Valley province is divided into the piedmont plain of the western San Joaquin Valley and the flood basin of the San Joaquin River.

Several of the larger streams draining the Diablo Range, including Little Panoche Creek, San Luis Creek, and Orestimba Creek, are within the study area. This study concentrates on the deposits from these and smaller streams that have accumulated on the piedmont plain of the western San Joaquin Valley as well as in local valleys within the foothills. These streams rarely reach the San Joaquin River even during floods, and thus their alluvial fans serve as local base level. The absence of Quaternary marine deposits in the valley at this latitude (Croft, 1972; Lettis, 1982) indicates that Quaternary sea-level fluctuations probably did not influence deposition of the alluvial deposits.

Climate and Vegetation

The west-central San Joaquin Valley is situated in the rain shadow of the Diablo Range and is characterized by a climate of long, hot, dry summers and mild winters. Figure 2 shows the monthly temperature and precipitation ranges for five weather stations in the San Joaquin Valley and Diablo Range. Stations at Newman, Los Banos, and Hanford show the distribution of temperature and precipitation from north to south along the axis of the valley; stations at Los Banos, San Luis Dam, and Idria show the change in climate from east to west with increasing elevation in the Diablo Range. In general, the climate ranges from temperate to semiarid in the northern San Joaquin Valley and from semiarid to arid in the southern San Joaquin Valley.

Most of the precipitation in the area falls as rain in the winter months, primarily from cyclonic storms that move eastward across California from the northeastern Pacific. In general, the annual precipitation increases with increasing elevation (15 to 20 cm in the valley, 20 to 38 cm in the foothills, and 38 to 51 cm in the central Diablo Range) and decreases from north to south within the valley and foothills (20 to 51 cm in the north, 15 to 25 cm in the south). Except for rare summer thunderstorms, intensity of the precipitation is low, and essentially all rainfall infiltrates the soil mantle.

Summers are warm to hot; typical mean daily maximum temperatures for July range from 33° to 38°C. In contrast, winters are cool; typical mean daily temperatures for January range from 0.5° to 3°C. In general, temperature increases from north to south and is inversely related to rainfall; temperatures are higher in the valley and lower in the central Diablo Range.

The distribution, density, and type of vegetation in the San Joaquin Valley and foothills of the Diablo Range are controlled strongly by climate and altitude. On the cooler, wetter slopes of the foothills, the native vegetation consists primarily of scattered oak and digger pine with an understory of sagebrush and various species of bunchgrass. On the warmer, drier piedmont plain and valley flood basin, bunchgrass and several varieties of herbaceous plants dominate the vegetation. In general, for the same altitude, the density of vegetation decreases from north to south in conjunction with decreasing precipitation and increasing temperature. Most of the native vegetation in these areas, however, has been removed or greatly altered by agriculture and by livestock grazing. Introduced annual grasses have largely replaced the native perennial bunchgrass over most of the flood basin, alluvial fans, and low foothill areas.

UPPER CENOZOIC DEPOSITS

Introduction and Depositional Environment

The upper Cenozoic deposits in the west-central San Joaquin Valley and adjacent foothills of the Diablo Range consist of weakly consolidated to unconsolidated gravel, sand, silt, and minor clay. The great range in grain size and texture of these deposits and their variety of sedimentary structures suggest that they accumulated in several depositional environments: alluvial fans deposited by fluvial and mudflow processes on the piedmont alluvial plain; riverine, flood-basin, paludal, and lacustrine deposits on the valley flood basin; stream-channel and flood-plain deposits on erosional surfaces in the foothills; and, locally, eolian deposits on the piedmont plain and valley flood basin. A detailed description of these facies and depositional environments appears in Lettis (1982).

Fluvial and mudflow deposits accumulating on the piedmont plain of the western San Joaquin Valley are composed of detritus derived entirely from the Diablo Range. These rocks include the Upper Jurassic and Cretaceous Franciscan assemblage, the Upper Jurassic, Cretaceous, and lower Tertiary Great Valley sequence, and a younger less extensive Tertiary sequence of marine and continental deposits. Figure 3 illustrates the distribution of these units in the study area.

Riverine, flood-basin, paludal, and lacustrine deposits on the valley flood basin are composed largely of detritus shed from the plutonic and metamorphic basement complex of the Sierra Nevada. These deposits typically are arkosic and were laid down principally by the San Joaquin River and associated sloughs, ponds, and oxbow lakes. Near the confluence of the San Joaquin River with Diablo Range streams, an admixture of varicolored lithic fragments derived from the Diablo Range is common.

The eolian sand is composed primarily of quartz and minor amounts of plagioclase and mica; lithic fragments from the Diablo Range are typically minor or absent. The sand is thus interpreted to be wind-reworked glacial-outwash deposits from the Sierra Nevada blown from the flood basin or eastern San Joaquin Valley onto the piedmont plain of the western San Joaquin Valley (Lettis, 1982).

The stratigraphic divisions recognized in this report commonly encompass two or more of these depositional facies. Figure 4 illustrates the interpreted distribution of facies within a single stratigraphic unit. From the flood basin to the mountain front, each unit generally consists of fine- to coarse-grained flood-basin and riverine deposits, locally interfingered with and overlain by fine-grained laminated middle- and lower-alluvial-fan deposits, overlain and commonly channeled by upper-fan coarse sand and gravel. Locally, eolian sand interfingers with and underlies the lower-fan deposits, and mudflow deposits interfinger with and overlie the upper-fan deposits. Facies relations between the eolian sand and flood-basin deposits have not been observed.

The complete sequence of facies is not exposed in any single outcrop; thus, Figure 4 is a composite of many isolated exposures supplemented by auger borings. The generalized vertical and lateral facies relations, however, suggest that each alluvial unit represents a progradation of alluvial fans from the Coast Ranges over the San Joaquin Valley flood basin, accompanied by and succeeded by mudflow deposition near the west margin of the valley.

Mapping Criteria

The upper Cenozoic deposits are divided into stratigraphic units and mapped primarily on the basis of relative ages inferred from geomorphic and pedologic criteria, as diagrammatically illustrated in Figure 5. Geomorphic criteria include: (1) relative topographic position in a sequence of inset stream terraces or nested alluvial fans; (2) truncation or incision of one alluvial fill by another; and (3) relative degree of surface modification, including erosional dissection, development of gilgai microrelief from the expansion and contraction of montmorillonitic soils, and biologic modification, such as the development of mima-mound microrelief (Arkley and Brown, 1954). Pedologic criteria include: (1) contrasting degrees of soil-profile development on depositional surfaces under similar conditions of parent material, relief, climate, vegetation, and drainage; and (2) crosscutting soil patterns resulting from a younger deposit and its characteristic pedogenic soil, overlying or truncating an older deposit characterized by a better developed pedogenic soil. In addition, stratigraphic superposition, as indicated by buried soils or erosional unconformities, becomes an important mapping criterion away

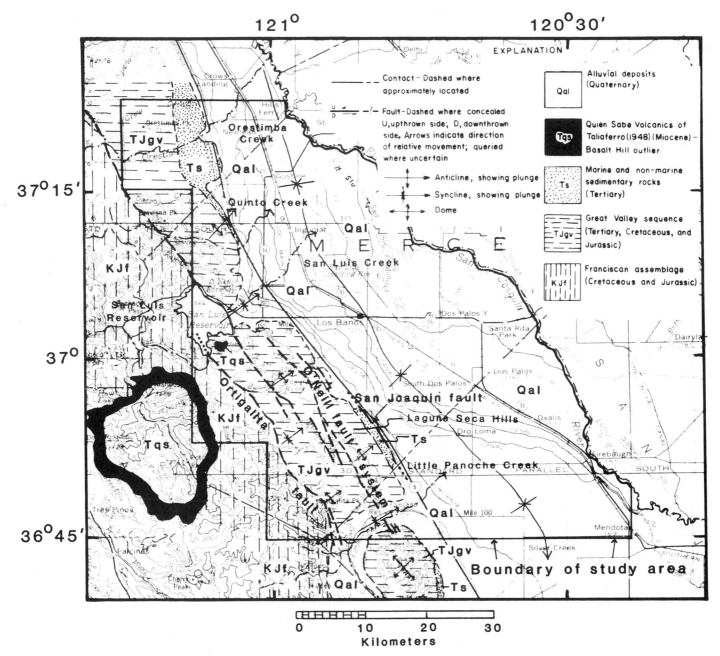

Figure 3. General distribution of pre-upper Cenozoic rocks in west-central San Joaquin Valley and bordering Diablo Range. Base from U.S. Geological Survey, 1:500,000 Topographic Series, California sheet.

from the mountain front, where coalesced, younger fan deposits commonly overlie coalesced older fan deposits.

These mapping criteria do not conform to the guidelines for recognition of formal rock-stratigraphic, biostratigraphic, or chronostratigraphic divisions (Cohee, 1974; Hedberg, 1976). Lateral and vertical lithologic variations within a single unit are frequently more pronounced than the differences between units, so the units are definable only to a limited degree by lithology. Absolute-age control and fossils generally are scarce in the deposits, so documented chronostratigraphic or biostratigraphic units

are not possible. Consequently, except for the previously defined Tulare Formation, informal names are used here for the stratigraphic divisions on the basis of the weathering and geomorphic criteria described above.

Stratigraphic Divisions

The upper Cenozoic deposits in the west-central San Joaquin Valley and adjacent foothills of the Diablo Range are divided into five stratigraphic units (Lettis, 1982), including, in

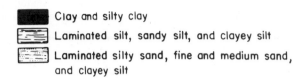

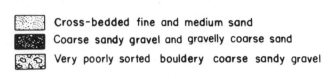

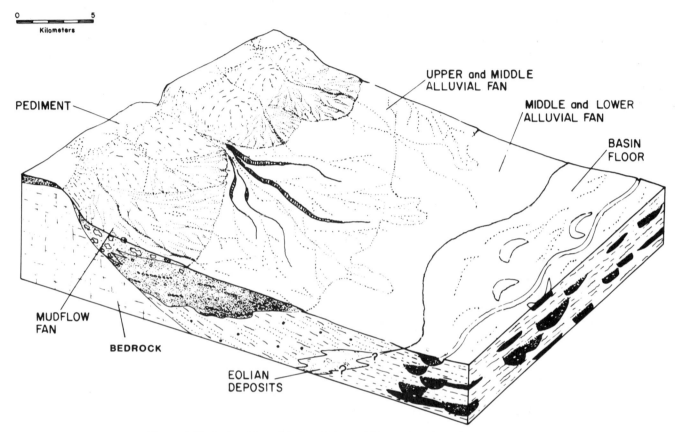

Figure 4. Vertical and lateral facies relations within one stratigraphic unit.

order of decreasing age: the Tulare Formation (Watts, 1894; Anderson, 1905) of late Pliocene and Pleistocene age, the informally named alluvium of Los Banos of middle and late Pleistocene age, the alluvium of San Luis Ranch of late Pleistocene and early Holocene age, and the alluvium of Patterson and the alluvium of Dos Palos, both of Holocene age. Tables 1 and 2 list the available absolute-age control. Tables 3 and 4 list pedologic characteristics of the late Cenozoic deposits. Detailed descriptions of the stratigraphic units and additional paleontologic, paleomagnetic, and pedologic age control are given by Lettis (1982).

These units are further divided into several informal and formal members. Primarily on the basis of geomorphic and pedologic criteria, the alluvium of Los Banos and the alluvium of San Luis Ranch are divided into three and two informal members, respectively; each member ranges in thickness from less than 1 up to 15 m and thus represents, at least in part, a distinct aggradational unit. Davis (1958) included in the subsurface of the San Joaquin Valley the 615,000- to 730,000-year-old lacustrine Cor-

coran Clay (Frink and Kues, 1954; Janda, 1965) as a formal lithologic member of the Tulare Formation.

Together, these upper Cenozoic deposits underlie all of the west-central San Joaquin Valley and large parts of the bordering foothills of the Diablo Range (Fig. 6). The Tulare Formation and the alluvium of Los Banos typically form thick deposits within stream valleys of the foothills and gravel veneers on extensive dissected pediments carved across the crest of the foothills. Younger terraces of the alluvium of San Luis Ranch and the alluvium of Patterson commonly are inset into the older deposits along streams draining the central Diablo Range (Fig. 7). Valleyward, each of these pediment and terrace deposits opens onto alluvial fans in the western San Joaquin Valley. Each alluvial fan commonly spills out over the next older fan, such that most of the valley floor is veneered by upper Pleistocene and Holocene deposits. Older alluvial fans composed of the alluvium of Los Banos and the oldest part of the alluvium of San Luis Ranch crop out only near the valley margin, where younger alluvial fans occupy

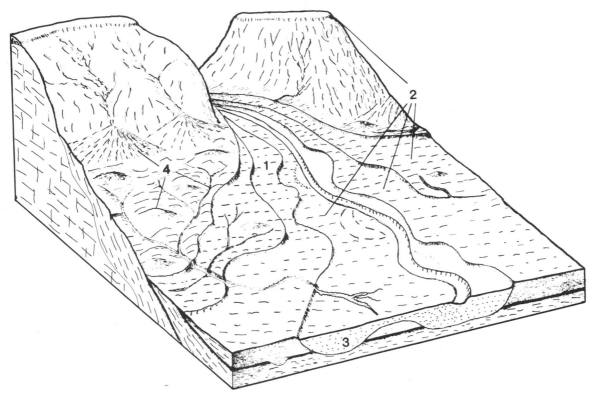

Figure 5. Geomorphology of upper Cenozoic deposits in west-central San Joaquin Valley. Along streams and near valley margin deposits form a sequence of inset stream terraces and nested alluvial fans; farther downslope, geomorphic surfaces merge, and deposits form coalesced and overlapping alluvial fans. Relative age criteria used to divide and map these deposits include: (1) relative topographic position in a sequence of inset alluvial fans or stream terraces; (2) relative degree of soil-profile development, which increases from little or no development on deposits adjacent to the trunk stream to very strong development on high terraces and pediment surfaces; (3) superposition in a vertical sequence indicated by buried soils or erosional unconformities; and (4) relative degree of surface modification, including development of microrelief, erosional dissection, and subsequent deposition.

surfaces incised into older alluvial-fan deposits. The Holocene alluvium of Dos Palos, which is confined to the flood basin of the San Joaquin Valley, consists of arkosic overbank and channel deposits of the San Joaquin River and associated sloughs.

Geomorphic evidence indicates that relative age is most useful within the foothills and near the valley margin, where cut-and-fill relations and inset surfaces are common. In the San Joaquin Valley, the degree of soil development and the superposition of deposits separated by buried soils are the primary criteria for distinguishing stratigraphic units (Fig. 5).

The degree of soil development on uneroded depositional surfaces typically increases with age, from little or no development on Holocene deposits to thick argillic or calcic profiles on middle Pleistocene deposits. Table 3 compares selected diagnostic properties of the stratigraphically important soils. In general, soil-profile thickness, thickness of the B_t horizon, degree (stage) of carbonate accumulation (Gile and others, 1965; Bachman and Machette, 1977), ped development, clay accumulation, and soil redness (Munsell) all increase with age. Figure 8 shows the typical distribution of upper Cenozoic deposits within the foothills and the relation of soils to the stratigraphic units.

LATE CENOZOIC STRUCTURE

Introduction

Quaternary deformation in the San Joaquin Valley and foothills of the Diablo Range has created a series of northeast- and northwest-trending broad gentle folds, which have been displaced vertically and laterally by three northwest-trending fault systems. These structures (Fig. 3) are manifest principally by tilted, warped, and offset Pliocene and Pleistocene terraces, pediments, and deposits; very few of the structures are shown by bedding attitudes or offset in underlying bedrock.

Folding

Quaternary folds include: (1) a large asymmetric syncline, trending N. 30°–40°W., underlying the San Joaquin Valley, and (2) many smaller northeast-trending anticlines and synclines, evident in the foothills, that may project into and slightly deform the San Joaquin Valley.

The San Joaquin Valley syncline governs the general posi-

TABLE 1. ^{14}C AGES OF UPPER CENOZOIC DEPOSITS IN THE
WEST-CENTRAL SAN JOAQUIN VALLEY, CALIFORNIA

Formation	Age (yr)	Latitude/Longitude north / west		Depth (m)	Remarks	References (Lab. No.)
Alluvium of Patterson	2850±100	36°35'30"	120°49'00"	4	Detrital charcoal; gives maximum age for aggradation of fill in Panoche Valley	Lettis (1982) (Beta 2345)
Alluvium of Patterson	2415±190	37°06'00"	121°02'00"	0.5	Gastropod shell: dates Holocene terrace along San Luis Creek	Lettis (1982) (Beta 1602)
Alluvium of Patterson	200±70	36°35'40"	120°46'00"	3.0-3.5	Detrital charcoal from laminated silt; gives maximum age for overlying fine to coarse sd.	Lettis (1982) (Beta 2616)
Alluvium of Dos Palos	3330±60	37°32'06"	121°07'05"	6.2	Detrital wood from contact of flood-basin slt. over arkosic sd.	B. F. Atwater and W. R. Lettis, unpub. data, 1981 (Beta 2788)
Alluvium of Dos Palos	1110±50	37°32'11"	121°06'35"	3.0-5.0	Detrital charcoal; crossbedded, arkosic sd., E. bank San Joaquin R.	do. (Beta 2786)
Alluvium of Dos Palos	8230±80	37°33'30"	121°07'02"	5.5	Detrital (?) wood; contact of channel gravel containing Coast Range detritus & overlying San Joaquin River arkosic sd. and slt.	do. (Beta 2787)
Alluvium of San Luis Ranch (upper member)	22,670±200	36°45'22"	120°18'56"	11.0	Wood; contact of arkosic sand & arkosic sand; max. limiting age of floodplain slt.; no pedogenesis	W. R. Lettis, 1982 (USGS 1201)
Alluvium of San Luis Ranch	28,200±330	36°44'50"	120°19'39"	15.0	Detrital wood; top of fining-upward sequence	do. (USGS 1197)
Alluvium of San Luis Ranch	29,970±250	36°44'50"	120°19'39"	15.0	Disseminated organic matter in slt. Top of fining-upward sequence	do. (USGS 1198)
Alluvium of San Luis Ranch	31,300±650	36°43'45"	120°21'53"	17.0	Detrital plant fragments from arkosic slt. & sd.; max. limiting age of buried surface (soil?)	do. (USGS 1200)
Alluvium of San Luis Ranch (lower member)	43,800 +1700-1400	36°43'45"	120°21'53"	17.0	Detrital wood in arkosic slt. & sd. overlain by fn. sd. from Coast Ranges (USGS 1200 dates contact)	do. (USGS 1199)

tion and orientation of the valley. This structure exists primarily because of westward tilting of the Sierran block in response to westward compression from the expanding Basin and Range province and uplift of the Coast Ranges. It has persisted since at least the Miocene, when the Coast Ranges were elevated, eroded, and overlain by the late Miocene Quien Sabe Volcanics of Taliaferro (1948).

The syncline is best expressed by structure contours on the surface of the Corcoran Clay Member of the Tulare Formation (Fig. 9; see also Croft, 1972, and Miller and others, 1971, for contour maps of the southern San Joaquin Valley). The clay defines an asymmetric trough whose western limb dips 1°–30° NE. and whose eastern limb dips less than 1° SW. In the northern and central San Joaquin Valley, the principal line of subsidence lies near the west margin, generally from 20 to 30 km west of the present courses of the San Joaquin River and Fresno Slough in the topographic axis of the valley. Subsurface data indicate that

rarely have the structural and topographic axes coincided during the Pleistocene (Lettis, 1982), suggesting that the average rate of deposition from the surrounding mountain ranges has equaled or exceeded the average rate of subsidence.

In the foothills, Quaternary pediments and alluvium are deformed into several broad arches and broad synclinal basins that include the arches over the Panoche, Wisenor, Los Banos, and Laguna Seca Hills, and an unnamed arch north of Quinto Creek. Structural basins include the Little Panoche, San Luis, and Salt Valleys, Carrisalito Flat, and, south of the study area, the Panoche Valley. Figure 10, which covers the Laguna Seca Ranch, Ortigalita Peak, Ortigalita Peak NW., and Charleston School 7.5-minute quadrangles (Fig. 1), shows a surface reconstruction of two elevated foothill pediments veneered by coarse sandy gravel of the alluvium of Los Banos and illustrates the pattern and magnitude of many of these folds.

If the structures have any significant direction of elongation,

TABLE 2. URANIUM-SERIES[1] AND POTASSIUM/ARGON AGES OF UPPER CENOZOIC DEPOSITS IN THE
WEST-CENTRAL SAN JOAQUIN VALLEY, CALIFORNIA

Formation	Age (yr.)	Latitude/Longitude north	west	Depth (m)	Remarks	References (Lab. No.)
Alluvium of San Luis Ranch (upper member)	16,000 (prelim.)	37°21'37"	121°10'30"	4.0-6.0	Equus bone; S. bank, Cow Creek; Underlies buried soil capped by Holocene deposits	E. L. Begg, U. C. Davis Personal Commun., 1981
Alluvium of Los Banos (upper member)	81,661±2000	36°37'15"	120°34'00"	15.0	Equus tooth fragments; U.S. Bur. of Rec. core hole 15/13-16N; above soil(?) separating alluvium of Los Banos from alluvium of San Luis Ranch	do.
Alluvium of Los Banos (upper member)	96,167±1623	37°43'30"	120°31'23"	?	Bison bone fragments; depth unknown; Veneer of alluvium of Los Banos in channel in underlying Tulare Fm.	do.
Alluvium of. Los Banos (upper member)	112,000±14,000	37°03'37"	121°01'25"	2.0	Groundwater carbonate concretion from fracture in clayey silt near O'Neill Forebay; max. limiting age for overlying veneer of alluvium of Los Banos	U.S.B.R., O'Neill Forebay trench study, 1980 prelim. rpt.
Alluvium of. Los Banos	160,000±5000	37°03'55"	121°01'30"	Surf.	Groundwater calcareous crust at surface of clayey silt unit; max. limiting age for overlying veneer of alluvium of Los Banos	do.
Tulare Fm.	250,000	36°45'35"	120°49'30"	12-15	Equus bone fragments; upper beds of upper part of Tulare Fm.	E. L. Begg, U. C. Davis, Personal commun., 1981
Turlock Lake Fm., Friant pumice Member	615,000±31,000	---	---	---	Sanidine from pumice; unit overlies Corcoran clay in sub-surface of the San Joaquin Valley. (This is the only K-Ar date for this study)	(Dalrymple, 1980) (Janda, 1965)

[1]Uranium-thorium disequilibrium dating of bone requires burial and subsequent incorporation of uranium transported in solution. Because thorium is relatively insoluble and not easily transported, dates are minimum ages. In arid and semiarid regions lacking near-surface ground water, slower migration of uranium in meteoric waters results in a mean residence time of uranium within the fossil that is less than the total elapsed time (Hansen and Begg, 1970).

it is approximately N. 40°–90° E., although the Panoche Hills arch seems to be a simple dome. The structures range from 5 to 20 km across and have structural relief ranging from 10 to 150 m. The Tulare Formation crops out on their flanks and dips gently to moderately from less than 5° to 20° NW. or SE.

The attitude of underlying Cretaceous and Tertiary beds, which strike N. 35°–50° W. and dip 30°–85° NE., typically persists with little change across the Quaternary structures. The single exception is the Los Banos Hills arch (Fig. 10), which has caused a 25° to 35° change in the strike of beds in the underlying Great Valley sequence (Briggs, 1953); this change indicates that deformation in this area probably began before Quaternary time. Deformation on all the other foothill structures may have begun during the Pleistocene.

The origin of these structures is poorly understood. Briggs (1953) recognized the Wisenor Hills and Los Banos Hills anticlines and suggested that the Los Banos Hills anticline reflects a local intrusion, similar in age to the Quien Sabe Volcanics, which has not been exposed by subsequent erosion. The Quien Sabe Volcanics, however, were erupted more than 7.5 m.y. B.P. (Snyder and Dickinson, 1979; Garniss Curtis, written commun., 1982), and unequivocal Quaternary deformation of the alluvium of Los Banos over the arch requires an alternative tectonic origin. Similarly, the other foothill arches cannot be attributed to high-level volcanic intrusion. The general absence of bedrock deformation associated with the Quaternary arches and basins suggests that these structures probably have been formed by flexural slip along bedding planes in the bedrock in response to continued flexural uplift and tilting of the foothill belt (see next subsection). Buckling under northwest-southeast-directed compression is unlikely in view of the general absence of folds in the underlying bedrock. However, the fold axes are nearly parallel to the least compressive stress axis of focal-plane solutions computed for the San Andreas fault system (including the Ortigalita fault) (La-Forge and Lee, 1982), suggesting that these folds may represent a response to stress along the San Andreas and are simply too broad and gentle to be manifest in the underlying steeply dipping bedrock. Warping due to extension along en echelon faults (i.e., pullapart basins) would create basins but not the broad, uplifted arches that are so prevalent in the foothills south of the San Luis Reservoir.

Faulting

Quaternary deposits are offset by three northwest-trending fault systems, including: (1) the Ortigalita fault system, which forms the principal contact between the Franciscan assemblage and the Great Valley sequence; (2) the San Joaquin fault system

TABLE 3. SOIL-HORIZON THICKNESS AND COLOR ON UPPER CENOZOIC ALLUVIAL DEPOSITS, WEST-CENTRAL SAN JOAQUIN VALLEY, CALIFORNIA

Parent material	Alluvial units on which soil is formed	Approximate depth to C horizon (cm)	Mean thickness of B horizon (cm)	Mean thickness of Bt horizon (cm)	Representative continuous moist subsoil color
	Patterson	0.0-0.6	0.0	0.0	2.5Y 5/2 to 10YR 5/3
	San Luis Ranch, upper member	0.3-1.0	0.0-60.0	0.0-40.0	10YR 3/2 to 10YR 5/3
Coarse-grained alluvial fan and terrace deposits	San Luis Ranch, lower member	0.8-1.7	20-115	20-60	10YR 4/3 to 7.5YR 4/4
	Los Banos, upper member	1.0-2.4	25-130	25-100	7.5YR 4/3 to 5YR 3/4
	Los Banos, middle and lower members	1.5-3.0	25-200	25-150	7.5YR 4/4 to 5YR 3/6
	Tulare Fm. (relict)	1.4-4.0	100-300+	100-300+	5YR 4/4 to 2.5YR 3/5
	Tulare Fm. (residual)	0.4-1.5	0.0-20	0.0	10YR 4/2 to 10YR 4/4
Fine-grained lower alluvial fan deposits	Patterson	0.5-1.2	0.0	0.0	5Y 4/2 to 2.5Y 4/3
	San Luis Ranch, upper member	1.3-1.8	40-115	20-40	2.5Y 4/4 to 10YR 4/2
	Dos Palos (channel)	0.0-0.4	0.0	0.0	2.5Y 4/3 to 10YR 4/3
Flood-basin deposits (arkosic parent material)	Dos Palos (flood plain)	0.5-1.3	0.0-110	0.0-20	2.5Y 4/4 to 10YR 3/1
	Modesto Fm.	0.5-1.5	15-100	15-60	2.5Y 5/4 to 10YR 4/4

(Herd, 1979b), which separates the foothills physiographic area from the San Joaquin Valley floor; and (3) the O'Neill fault system, originally named by Herd (1979a) for a small bedding-plane fault near O'Neill Forebay, but which is extended here to include all the geologically similar small bedding-plane faults between the San Joaquin and Ortigalita fault systems (Figs. 3 and 10). Figure 11 shows the interpreted position and regional relations of these faults.

Geomorphic expressions of faulting that indicate late Quaternary tectonism include: (1) linear escarpments separating several levels of foothill pediments veneered by deposits of the alluvium of Los Banos; (2) lateral continuity of the escarpments across interfluve areas, which precludes their origin as fluvial-terrace escarpments; and (3) the presence of sag ponds, active springs, and drainage reversals, and the development of restricted foothill valleys along the fault trace.

Ortigalita Fault

The Ortigalita fault was first shown by Anderson and Pack (1915) to form the approximate contact between the Franciscan assemblage and Great Valley sequence (Fig. 11). Later workers, including Taliaferro (1943), Leith (1949), Briggs (1953), and Page (1980), described the fault in greater detail and suggested that reverse, thrust, or normal displacement along the Ortigalita

and related faults may have elevated the central Diablo Range, underlain by the Franciscan assemblage, to its present position during the Pliocene and, possibly, the Pleistocene.

The presence of roughly coeval basalt flows of the Quien Sabe Volcanics at similar elevations on both sides of the fault, however, constrains the timing of this activity. Basalt Hill is a small outlier of the Quien Sabe Volcanics east of the fault (southeast of San Luis Reservoir, Fig. 3), K-Ar dated as 9.35 ± 0.1 m.y. (Garniss Curtis, written commun., in Lettis, 1982). This undeformed outlier is interpreted as a remnant of a basalt flow from the main Quien Sabe volcanic field (Lettis, 1982). Occurrence of the base of the outlier at an elevation near that of the base of the main field suggests that significant uplift of the range relative to the foothills ceased before the late Miocene. Extension of vertically undisplaced embayments of the Tulare Formation and the alluvium of Los Banos into the central Diablo Range across the fault also indicates that little, if any, uplift of the central range relative to the foothills has occurred during the Quaternary.

Quaternary activity is indicated, however, by a conspicuous morphologic expression of the fault trace. Sag ponds, springs, and prominent, weakly dissected escarpments delineate the fault trace. The presence of these features in Tulare, Los Banos, and San Luis Ranch deposits suggests late Pleistocene activity. Local offset of the alluvium of Patterson, particularly in the headwater area of Little Panoche Creek, also suggests Holocene activity.

TABLE 4. STRUCTURAL PROPERTIES OF SOILS DEVELOPED ON UPPER CENOZOIC DEPOSITS,
WEST-CENTRAL SAN JOAQUIN VALLEY, CALIFORNIA

Parent material	Alluvial unit on which soil is formed	Typical carbonate accumulation	Clay and iron oxides in best developed subsoil horizon	Dry consistency of subsoil	Dry structure in best developed subsoil horizon
Coarse-grained alluvial fan and terrace deposits	Patterson	None to Stage I, disseminated, filament	None evident	Firm to slightly hard	Massive, granular or fine, subangular blocky
	San Luis Ranch (upper member)	Stages I and II, disseminated soft masses; nodular	None to thin discontinuous coatings on ped facings	Slightly hard to hard	Massive, granular or weak, coarse, subangular blocky
	San Luis Ranch (lower member)	Stage II, soft masses; nodular	Moderate to thin, discontinuous to nearly continuous coatings on ped faces and large clasts; pores partially filled	Hard to very hard	Moderate fine to medium subangular blocky and weak, fine prismatic
	Los Banos (upper member)	Stages II and III, nodular, massive	Common thick continuous coatings on ped faces and clasts; pores partially filled	Hard to very hard	Weak coarse prismatic, breaking into medium to coarse subangular blocky
	Los Banos (middle and lower members)	Stages II and III, nodular, massive	Common thick continuous coatings on ped faces and clasts; pores nearly filled	Hard to very hard	Strong coarse prismatic breaking into angular blocks
	Tulare Fm. (relict)	Stages I to IV disseminated to laminar depending on topography	Thick continuous coatings on peds and clasts; pores nearly filled or filled; slickensides common	Hard to very hard	Strong coarse prismatic breaking into subangular or angular blocky
	Tulare Fm. (residual	Stages I to II disseminated, nodular	None to thin discontinuous coatings on peds and clasts	Slightly hard to hard	Weak fine prismatic breaking into weak fine subangular blocky and granular
Fine-grained lower alluvial fan deposits	Patterson	None to Stage I disseminated, rare nodules	None evident	Very hard	Strong, coarse prismatic and subangular blocky in lower A horizon
	San Luis Ranch (upper member)	Stage II, soft nodular masses	Thin, continuous on peds, pores partially filled	Very hard	Moderate medium subangular blocky and moderate medium prismatic
	Dos Palos (channel)	None to Stage I,	None evident	Slightly hard	Weak fine granular
Flood-basin deposits (arkosic origin)	Dos Palos (flood plain)	Stages I and II, disseminated, nodular soft masses	None to thin discontinuous coatings on peds and pores	Hard	Weak medium subangular blocky to moderate coarse prismatic
	Modesto Formation	Stages II and III nodular, massive	Thin continuous and discontinuous coatings on peds; pores partially filled	Hard to very hard	Moderate medium subangular blocky to strong coarse prismatic

The sense of Quaternary displacement, though not well understood, clearly differs from the Miocene and earlier vertical displacement which may have elevated the central Diablo Range. Focal-plane solutions computed for recent tremors (LaForge and Lee, 1982), trenching studies of the faults by the U.S. Bureau of Reclamation (Anderson and others, 1982), and the en echelon character of the fault trace all suggest that a significant component of lateral slip characterizes fault displacement in the area.

The extent of Quaternary lateral displacement is not known. The absence of conspicuous consistent morphologic offsets of ridges or creeks along the fault trace, however, suggests that any displacements exceeding several kilometers probably occurred before establishment of the present drainage network of the foothill belt during the late Pleistocene. The position of the Basalt Hill outlier of the Quien Sabe volcanic field east of the fault and north of the main volcanic field (Fig. 3) indicates that significant late

Cenozoic right-lateral displacement of tens of kilometers along the Ortigalita fault is not realistic.

San Joaquin Fault System

The San Joaquin fault system, situated along the foothill-valley margin, in many areas produces a prominent linear east-facing faceted escarpment (Fig. 11). The escarpment is especially well preserved along the front of the Laguna Seca Hills north of Little Panoche Creek (Figs. 3 and 10).

The orientation of the fault plane and the sense of Quaternary displacement are poorly understood. Vertical displacement, reported by Herd (1979b) along an east-dipping fault plane north of the study area near Ingram Creek, is indicated by vertical truncation of the reconstructed pediment surfaces veneered by the alluvium of Los Banos along the front of the Laguna Seca Hills

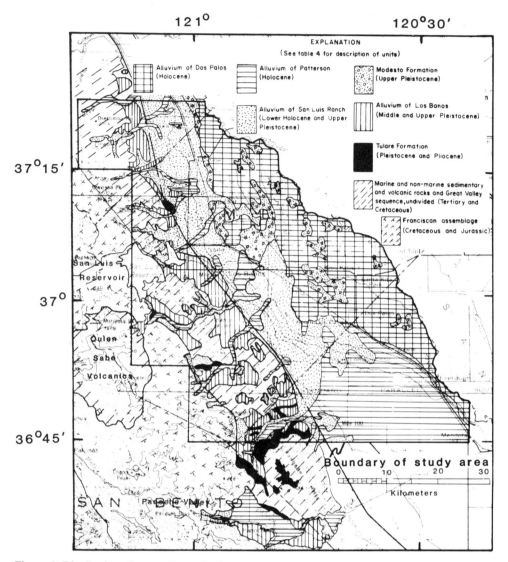

Figure 6. Distribution of upper Cenozoic deposits in west-central San Joaquin Valley and bordering foothills of Diablo Range. Base from U.S. Geological Survey 1:500,000 Topographic Series, California sheet.

(Fig. 10). In the Laguna Seca Hills, the middle member of the alluvium of Los Banos veneers a pediment, at one time graded to the valley floor, that now projects approximately 140 m above the present valley floor. This displacement provides an average rate of vertical offset in this area of about 45 to 65 cm per millenium, using the minimum and maximum probable age of 200,000 to 300,000 years for the pediment. However, the fault trace commonly is overlain by unfaulted alluvium of San Luis Ranch and alluvium of Patterson across the fan heads of the larger Diablo Range drainages, a relation indicating little activity along the fault during at least the past 40,000 to 60,000 years.

O'Neill Fault System

The O'Neill fault system includes the zone of numerous small foothill faults between and subparallel to the larger Ortigal-

ita and San Joaquin faults (Figs. 3, 10, and 11). These faults trend N. 35°–50°W., parallel to the strike of beds in the Cretaceous and Tertiary bedrock, and are evident principally between Little Panoche Creek on the south and San Luis Creek on the north. The faults differ from the larger, bordering faults in that they appear to exhibit predominantly bedding-plane slip in response to continued uplift and tilting of the foothill belt during the Quaternary.

Fault displacement is manifest principally by offset of the broad foothill pediments veneered by the alluvium of Los Banos. Contour reconstruction of these surfaces suggests that vertical displacement ranges from less than 1 to 100 m (Fig. 10). The broad Quaternary folds shown in Figure 10, however, are not displaced laterally, and other evidence of lateral displacement has not been observed.

The magnitude of vertical displacement changes not only between individual fault segments but also laterally along each

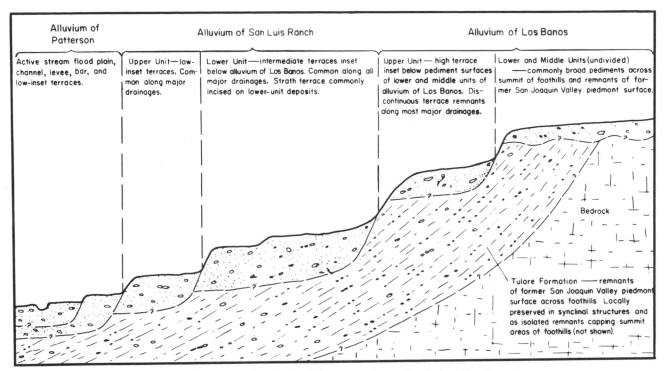

Alluvium of Patterson	Alluvium of San Luis Ranch		Alluvium of Los Banos	
Active stream flood plain, channel, levee, bar, and low-inset terraces.	Upper Unit—low-inset terraces. Common along major drainages.	Lower Unit—intermediate terraces inset below alluvium of Los Banos. Common along all major drainages. Strath terrace commonly incised on lower-unit deposits.	Upper Unit—high terrace inset below pediment surfaces of lower and middle units of alluvium of Los Banos. Discontinuous terrace remnants along most major drainages.	Lower and Middle Units (undivided) —commonly broad pediments across summit of foothills and remnants of former San Joaquin Valley piedmont surface.

Bedrock

Tulare Formation —remnants of former San Joaquin Valley piedmont surface across foothills. Locally preserved in synclinal structures and as isolated remnants capping summit areas of foothills (not shown).

Figure 7. Diagrammatic cross section of a foothill stream valley and bordering hills, illustrating geomorphology of stratigraphic units within foothills of Diablo Range.

segment. Maximum vertical offset typically coincides with maximum anticlinal flexure; displacement decreases down the flanks to less than 5 m within most of the synclinal basins.

Available field data suggest that the principal sense of motion in the O'Neill fault system is reverse displacement along east-dipping fault planes. Fault escarpments have a consistent west-facing aspect, and where the escarpments are traced into the homoclinally east-tilted Cretaceous and Tertiary bedrock, they appear only as sheared shale units between more resistant sandstone beds. Bedding is not displaced.

The apparent reverse displacement on the faults and the absence of disruption in underlying bedrock suggest a sympathetic bedding-plane-slip origin in response to Quaternary flexural folding of the foothill region and (or) fault activity along the larger Ortigalita and San Joaquin faults. The coincidence of maximum displacement with local arching described above suggests, in turn, that the small foothill arches and basins may be a result of bedding-plane slippage.

DEPOSITIONAL MODEL

The upper Cenozoic alluvial deposits and landforms in the west-central San Joaquin Valley and adjacent foothills of the Diablo Range record periods of both landscape instability, accompanied by erosion in the foothills and alluviation in the valley, and landscape stability, when soils developed on stable surfaces. Hypothetical causes for these periods of alternating regional landscape stability and instability fall into three principal categories, or some combination of these categories: (1) rise in

base level along the lower San Joaquin River, particularly in response to a rise in sea level; (2) episodic uplift of the foothills or central Diablo Range relative to the San Joaquin Valley; and (3) climatic fluctuations from more humid to more arid conditions. Direct field evidence and sufficient absolute-age control that might favor one of these models are not available in the western San Joaquin Valley. Furthermore, cut and fill by the lower San Joaquin River in the valley axis effectively prevents rock-stratigraphic correlation of the western San Joaquin Valley alluvial units with the climatically controlled glacial-outwash alluvial units of the eastern San Joaquin Valley described by Marchand and Allwardt (1981). Justification for a model, therefore, must rely on less direct evidence. Such evidence favors alluviation in response to climatic fluctuation, largely because it precludes alluviation in response to episodic uplift or base-level change.

The five informally designated subunits of the alluviums of Los Banos and San Luis Ranch indicate that the San Joaquin Valley and Diablo Range were subject to at least five periods of landscape instability separated by soil-forming intervals during the past 500,000 years. Recognition of these post-Tulare units over 1,800 km^2 of the western San Joaquin Valley (Fig. 6) suggests that they more likely reflect regional climatic fluctuations or changes in base level than local episodic tectonism. The deposits locally rest on or interfinger with wind-reworked glacial outwash from the eastern San Joaquin Valley, a relation suggesting that they are coeval with glacial-outwash units of the eastern San Joaquin Valley.

Climatically induced eustatic fluctuations in sea level hypothetically can cause alternating periods of aggradation and inci-

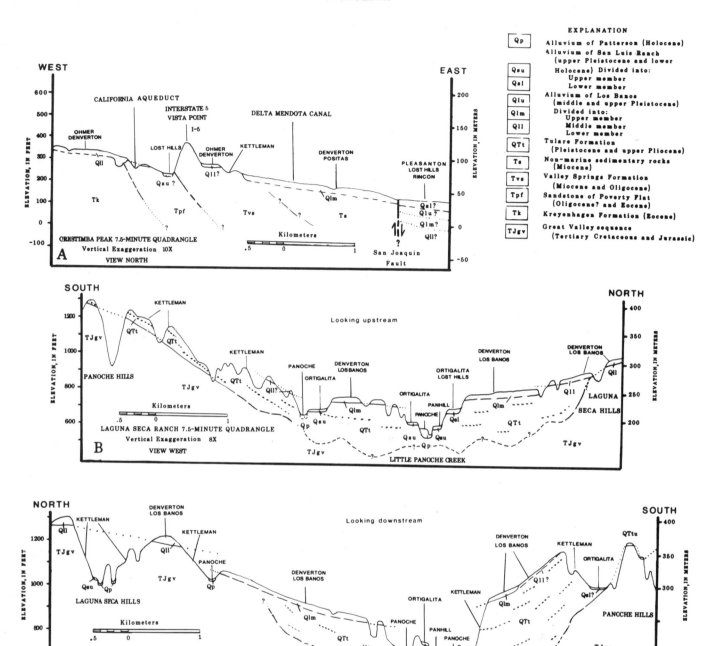

Figure 8. Detailed geologic sections near Orestimba and Little Panoche Creeks, showing lithology of upper Cenozoic deposits in foothills of Diablo Range and relation of soils to stratigraphic units and geomorphic surfaces. Soil-series names are in capitals above ground surface. Stratigraphic units are shown generally below ground surface. Cross sections through U.S. Geological Survey 7.5-minute quadrangles, Laguna Seca Ranch.

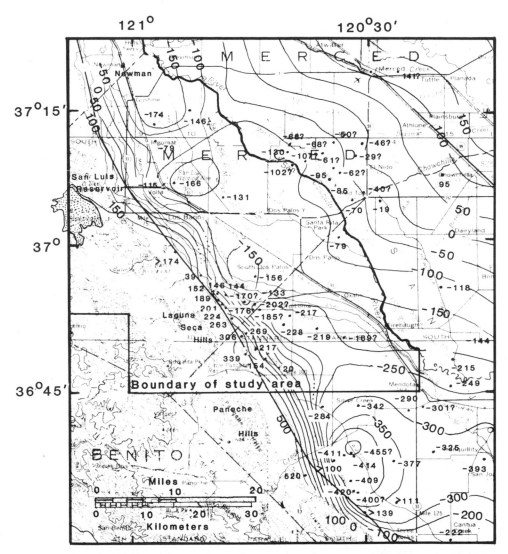

Figure 9. Structure-contour map of top of the Corcoran Clay Member of the Tulare and Turlock Formations in northern San Joaquin Valley, showing distribution and deformation of clay in north-central San Joaquin Valley. Contour interval, 50 ft, except east of Panoche Hills, where 100-ft contours are shown. Subsea depth to clay is listed adjacent to drill holes (dots) in valley; queried depths are from water wells. Base from U.S. Geological Survey 1:500,000 Topographic Series, California sheet.

sion along the major Diablo Range streams either by directly inundating the San Joaquin Valley with marine water or, more likely, by causing entrenchment or aggradation of the San Joaquin River and thereby indirectly changing local base level for the Diablo Range streams. Such fluctuations, however, probably did not generate the observed post-Tulare alluvial deposits: (1) Pleistocene marine deposits indicative of a marine transgression are absent in the central San Joaquin Valley, and (2) in the modern climate, streamflow from each of the Diablo Range streams in the study area is absorbed into its alluvial fan and rarely reaches the San Joaquin River. Therefore, each alluvial fan commonly serves as local base level for its parent stream, and the streams are not influenced by possible sea-level-induced changes on the channel gradient of the axial river.

It is probable that the pluvial period during which the Corcoran Clay accumulated caused higher local base levels during the middle Pleistocene, thus inducing alluvial-fan or deltaic aggradation of part of the Tulare Formation. Post-Tulare deposits, however, cannot be attributed to such base-level changes because, as argued by Frink and Kues (1954), the lake in which the Corcoran Clay was deposited appears to be unique in the Quaternary history of the San Joaquin Valley.

Considerable evidence argues against alluviation as a result of episodic uplift of the Diablo Range relative to the San Joaquin Valley. (1) Post-Tulare deposits occur along streams draining foothill areas of both Quaternary upwarping (e.g., Los Banos and Laguna Seca Creeks) and downwarping (e.g., Little Panoche and San Luis Creeks). (2) Most of the modern stream channels and

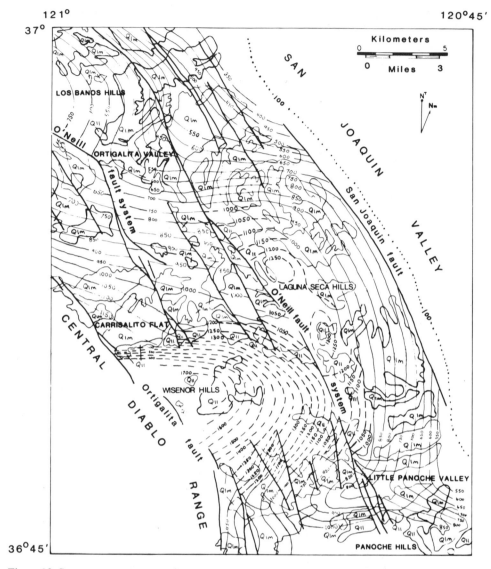

Figure 10. Structure-contour map of reconstructed pediment surfaces in foothills of Diablo Range. Solid contours drawn on middle member (Unit Q1m) of the alluvium of Los Banos; dashed contours drawn on lower member (Unit Q11) of the alluvium of Los Banos; dotted contours drawn on surface of the Corcoran Clay Member (of the Tulare Formation). Contour interval, 50 ft. Heavy solid lines, faults. Only generalized outcrops of pediments are shown; pediments are typically veneered by gravelly coarse sand of the alluvium of Los Banos. See Figure 1 for location.

alluvial units are inset into pre-existing alluvium at the foothill-valley margin, whereas incision into bedrock would be expected if uplift of the foothills induced alluviation. (3) The San Joaquin and O'Neill faults do not displace the alluvium of San Luis Ranch or Patterson and thus cannot be the cause of late Pleistocene alluviation. (4) Although the larger Diablo Range streams drain the central Diablo Range, underlain by the Franciscan assemblage, and detritus shed from the Franciscan makes up much of the alluvium in the western San Joaquin Valley, the central Diablo Range has not been significantly elevated relative to the foothills along the Ortigalita fault since the late Miocene.

On the basis of these lines of evidence, the alluvial units and their soils are here interpreted to result from climatic fluctuations during the Quaternary. The Diablo Range has not been glaciated, however, and the relation of these fluctuations to glacial and interglacial periods of the Quaternary is difficult to determine. It has long been observed and reported, however, that climatically induced changes in surface runoff and the type and density of vegetation in nonglaciated areas strongly influence the weathering, erosion, transport, and deposition of sediment (Gilbert, 1877; Huntington, 1907, 1914; Eckis, 1928; Cotton, 1945; Blissenbach, 1954; Schumm, 1965; Gile and others, 1981).

From these ideas, the following hypothetical model best accounts for the alluvial units in the western San Joaquin Valley, in contrast to the climatic model proposed for glacial outwash from the Sierra Nevada (Arkley, 1962; Janda, 1965; Marchand,

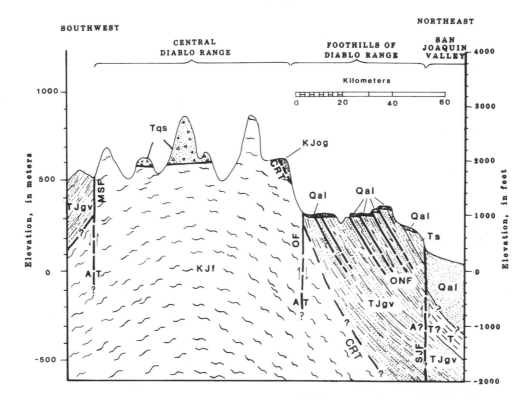

Figure 11. Generalized cross section of Diablo Range at approximately lat 36°50′N, illustrating pertinent lithologic and structural features of central Diablo Range and eastern foothills. Qal, upper Cenozoic alluvium; Tqs, Quien Sabe Volcanics of Taliaferro (1948); Ts, Tertiary marine and nonmarine sedimentary rocks; KJf, Franciscan assemblage; TJgv, Great Valley sequence; KJog, outlier of basal ophiolite and lower ends of the Great Valley sequence, undivided (shown diagrammatically); MSF, Madrone Springs and related faults; OF, Ortigalita fault; ONF, O'Neill fault system; SJF, San Joaquin fault system; CRT, Coast Range thrust. Faults dashed where approximately located; queried where uncertain. A, movement away from observer; T, movement toward observer. Arrows indicate direction of relative movement.

1977). Intervals of maximum erosion and resulting deposition coincide with transitions from the more humid conditions of glaciation to the more arid conditions of an interglacial period. A dense vegetation cover of grasses, shrubs, and trees, similar to that on the higher, wetter, sheltered slopes of the central Diablo Range, probably deteriorated during this transition to the scattered native shrub and bunchgrass vegetation covering the foothills today, and slope material weathered during the preceding humid period was exposed and became erodible. During maximum aridity of an interglacial period, fluvial deposition in the valley was hindered because of insufficient runoff, and mudflows were common. During transitions from arid to humid conditions, erosion and resulting deposition were hindered because of increased vegetation cover and because most of the slope material that weathered during the preceding humid period had been removed.

REFERENCES CITED

Anderson, F. M., 1905, A stratigraphic study in the Mount Diablo Range of California: California Academy of Science Proceedings, ser. 3, v. 2, p. 155–248.

Anderson, L. W., Anders, M. H., and Ostenaa, D. A., 1982, Late Quaternary faulting and seismic hazard potential, eastern Diablo Range, California [abs.] *in* Proceedings, Conference on earthquake hazards in the eastern San Francisco Bay Area: Hayward, California, p. 2.

Anderson, R., and Pack, R. W., 1915, Geology and oil resources of the west border of the San Joaquin Valley north of Coalinga, California: U.S. Geological Survey Bulletin 603, 220 p.

Arkley, R. J., 1962, The geology, geomorphology, and soils in the San Joaquin Valley in the vicinity of the Merced River, California, *in* Geologic guide to the Merced Canyon and Yosemite Valley: California Division of Mines and Geology Bulletin 182, p. 25–32.

Arkley, R. J., and Brown, H. C., 1954, The origin of Mima mound (hogwallow) microrelief in the far western states: Soil Science Society of America Proceedings, v. 18, no. 2, p. 195–199.

Bachman, G. O., and Machette, M. N., 1977, Calcic soils and calcretes in the southwestern United States: U.S. Geological Survey Open-File Report 77-794, 163 p.

Blissenbach, E., 1954, Geology of alluvial fans in semiarid regions: Geological Society of America Bulletin, v. 65, p. 175–189.

Briggs, L. I., Jr., 1953, Geology of the Ortigalita Peak quadrangle, California: California Division of Mines and Geology Bulletin 167, 61 p., scale

1:62,500.

Cohee, G. V., Chairman, 1974, Stratigraphic nomenclature in reports of the U.S. Geological Survey: Washington, D.C., U.S. Government Printing Office, 45 p.

Cotton, C. A., 1945, The significance of terraces due to climatic oscillation: Geological Magazine, v. 82, p. 10–16.

Croft, M. G., 1972, Subsurface geology of the late Tertiary and Quaternary water-bearing deposits of the southern part of the San Joaquin Valley, California: U.S. Geological Survey Water-Supply Paper 1999-H, p. H1–H29.

Dalrymple, G. B., 1980, K-Ar ages of the Friant Pumice member of the Turlock Lake Formation, the Bishop Tuff, and the Tuff of Reds Meadow, central California: Isochron/West, no. 28, p. 3–5.

Davis, G. H., 1958, Progress report on land-subsidence investigations in the San Joaquin Valley, California, through 1957: Inter-agency committee on land subsidence in the San Joaquin Valley, Sacramento, California, 160 p.

Davis, S. N., and Hall, F. R., 1958, Late Cenozoic history of the northeastern San Joaquin Valley, California [abs.]: Geological Society of America Bulletin, v. 69, no. 2, p. 1532.

Eckis, R., 1928, Alluvial fans in the Cucamonga district, southern California: Journal of Geology, v. 36, p. 224–247.

Frink, J. W., and Kues, H. A., 1954, Corcoran Clay, a Pleistocene lacustrine deposit in the San Joaquin Valley, California: American Association of Petroleum Geologists Bulletin, v. 38, p. 2357–2371.

Gale, H. S., Piper, A. M., and Thomas, H. A., 1939, Geology of the Mokelumne area, California, *in* Piper, A. M., and others, eds., Geology and ground-water hydrology of the Mokelumne area, California: U.S. Geological Survey Water-Supply Paper 780, p. 14–100.

Gilbert, G. K., 1877, Report on the geology of the Henry Mountains: Washington, D.C., U.S. Government Printing Office, 160 p.

Gile, L. H., Hawley, J. W., and Grossman, R. B., 1981, Soils and geomorphology in the Basin and Range area of southern New Mexico—Guidebook to the Desert Project: New Mexico Bureau of Mines and Mineral Resources Memoir 39, 222 p.

Gile, L. H., Peterson, F. F., and Grossman, R. B., 1965, The K horizon—a master soil horizon of carbonate accumulation: Soil Science, v. 99, no. 2, p. 74–82.

Hansen, R. O., and Begg, E. L., 1970, Age of Quaternary sediments and soils in the Sacramento area, California, by uranium and actinium series dating of vertebrate fossils: Earth and Planetary Science Letters, v. 8, p. 411–419.

Hedberg, H. D., ed., 1976, International stratigraphic guide: New York, John Wiley and Sons, 200 p.

Helley, E. J., 1967, Sediment transport in the Chowchilla River basin, Mariposa, Madera, and Merced Counties, California [Ph.D. thesis]: Berkeley, University of California, 153 p.

Herd, D. G., 1979a, Geologic map of O'Neill Forebay, western Merced County, California: U.S. Geological Survey Open-File Report 79-359, scale 1:24,000.

—— 1979b, The San Joaquin fault zone: Evidence for late Quaternary faulting along the west side of the northern San Joaquin Valley, California [abs.]: Geological Society of America Abstracts with Programs, v. 11, p. 83.

Huntington, E., 1907, Some characteristics of the glacial period in nonglaciated regions: Geological Society of America Bulletin, v. 18, p. 351–388.

—— 1914, The climatic theory of terraces, *in* Huntington, E., The climatic factor as illustrated in arid America: Carnegie Institution of Washington Publica-tion 192, p. 23–36.

Janda, R. J., 1965, Quaternary alluvium near Friant, California, *in* Wahrhaftig, C., and others, eds., Northern Great Basin and California: International Association for Quaternary Research (INQUA) Congress, 7th Guidebook for Field Conference I, p. 128–133.

—— 1966, Pleistocene history and hydrology of the upper San Joaquin River, California [Ph.D. thesis]: Berkeley, University of California, 425 p.

Jenkins, O. P., 1943, Geomorphic provinces of California, *in* Geologic formations and economic development of the oil and gas fields of California: California Division of Mines and Geology Bulletin 118, p. 83–88.

LaForge, R., and Lee, W.H.K., 1982, Seismicity and tectonics of the Ortigalita fault and southeast Diablo Range, California [abs.], *in* Proceedings, Conference on earthquake hazards in the eastern San Francisco Bay Area: Hayward, California, p. 26.

Leith, C. J., 1949, Geology of the Quien Sabe quadrangle, California: California Division of Mines and Geology Bulletin 147, 59 p., scale: 1:62,500.

Lettis, W. R., 1982, Late Cenozoic stratigraphy and structure of the western margin of the central San Joaquin Valley, California: U.S. Geological Survey Open-File Report 82-526, 203 p.

Marchand, D. E., 1977, The Cenozoic history of the San Joaquin Valley and adjacent Sierra Nevada as inferred from the geology and soils of the eastern San Joaquin Valley, *in* Singer, M. J., ed., Soil development, geomorphology, and Cenozoic history of the northeastern San Joaquin Valley and adjacent area, California: Joint Field Session, American Society for Agronomy, Soil Science Society of America, and Geological Society of America, Guidebook, Davis, University of California Press, p. 39–50.

Marchand, D. E., and Allwardt, A., 1981, Late Cenozoic stratigraphic units, northeastern San Joaquin Valley, California: U.S. Geological Survey Bulletin 1470, 70 p.

Miller, R. E., Green, J. H., and Davis, G. H., 1971, Geology of the compacting deposits in the Los Banos-Kettleman City subsidence area, California: U.S. Geological Survey Professional Paper 497-E, p. E1-E46.

Page, B. M., 1980, The southern Coast Ranges, *in* Ernst, W. G., ed., The geotectonic development of California (Rubey volume 1): Englewood Cliffs, New Jersey, Prentice-Hall, p. 329–417.

Schumm, S. A., 1965, Quaternary paleohydrology, *in* Wright, H. E., Jr., and Frey, D. G., eds., The Quaternary of the United States: Princeton, Princeton University Press, p. 783–794.

Snyder, W. S., and Dickinson, W. R., 1979, Geometry of triple junctions related to San Andreas transform: Journal of Geophysical Research, v. 84, p. 561–572.

Taliaferro, N. L., 1943, Geologic history and structure of the central Coast Ranges of California, *in* Geologic formations and economic development of the oil and gas fields of California: California Division of Mines and Geology Bulletin 118, p. 119–163.

—— 1948, Geologic map of the Hollister quadrangle, California: California Division of Mines and Geology Bulletin 143, no text, scale 1:62,500.

U.S. National Oceanic and Atmospheric Administration, 1967–1976, Climatological Data Bulletin, Volumes 71 through 80.

Watts, W. L., 1894, The gas and petroleum yielding formation of the Central Valley of California: California State Mining Bureau Bulletin 3, 100 p.

Manuscript Accepted by the Society January 12, 1985

Geological Society of America
Special Paper 203
1985

Correlation and age of Quaternary alluvial-fan sequences, Basin and Range province, southwestern United States

Gary E. Christenson
Utah Geological and Mineral Survey
Salt Lake City, Utah 84108

Charles (Rus) Purcell
3075 Molokai Place
Costa Mesa, California 92626

ABSTRACT

Alluvial-fan deposits in the Basin and Range province in Arizona, California, Nevada, and Utah may be correlated by relative-age criteria. Diagnostic criteria include (1) drainage pattern, (2) incision depth, (3) surface morphology, (4) desert pavement and varnish development, (5) soil-profile development, and (6) morphostratigraphic relations. These criteria allow us to group alluvial-fan deposits into three age classes: young, intermediate, and old. Young fans have a distributary drainage pattern (bar and channel topography), stream incision typically less than 1 m, and an undeveloped to weak soil profile. In contrast, intermediate-age fans have a dendritic to parallel drainage pattern, major channel incision of about 1 to 10 m with undissected interfluves, and weak to strong soil profiles. Old fans retain little of their original surface morphology, have stream incision greater than about 10 m, and typically are cut off from their original source areas by modern drainages. Soils are strongly developed on remnant drainage divides but elsewhere are generally removed by erosion.

Similar sequences of alluvial fans throughout the Basin and Range suggest a regional control over deposition probably related to Quaternary climatic changes. Local variations in sequences may result from other factors which influence deposition, such as lithology and tectonic or other base-level controls. However, fans attributable to these factors are generally not of significance in regional correlation. Relative soil-profile development, morphostratigraphic relations, and absolute dates from numerous Basin and Range localities indicate that most young fans are less than 15,000 years old; intermediate-age fans range from 10,000 to 700,000 years old; and old fans generally exceed 500,000 years in age.

INTRODUCTION

Broad piedmont fan systems cover large areas of the Basin and Range province of the southwestern United States. Each piedmont consists of a complex of coalescing alluvial fans that vary in size, composition, and age. These alluvial-fan deposits constitute a significant part of the Quaternary record in the Basin and Range province, and determining their age is important to reconstructing the Quaternary stratigraphic and tectonic history of the region.

Alluvial-fan deposits vary greatly in age, both within a single fan and between adjacent fans on an alluvial piedmont (summarized in Cooke and Warren, 1973). Several workers have developed similar qualitative techniques to differentiate alluvial-fan deposits according to relative age (see, for example, Denny, 1965; Bull, 1964, 1974b; Shlemon and Purcell, 1976; Shlemon, 1978; and Hoover and others, 1981). Diagnostic criteria used to differentiate alluvial-fan sequences are: drainage pattern, fan morphology, desert pavement and varnish development, soil-profile (pedogenic) development, and morphostratigraphic relations.

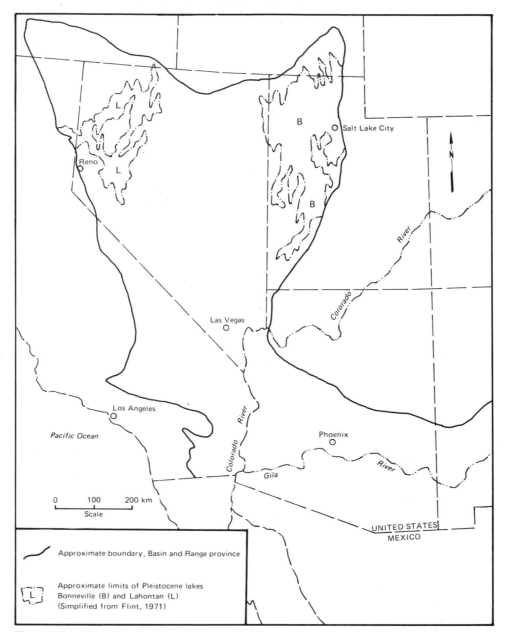

Figure 1. Location map showing the Basin and Range province and approximate limits of Pleistocene Lakes Bonneville and Lahontan (Flint, 1971).

Many workers in the southwestern United States (Bull, 1974a; Shlemon and Purcell, 1976; Shlemon, 1978; Hoover and others, 1981) have considered alluviation to be climatically controlled. The influence of climate, however, is difficult to recognize in correlating Basin and Range alluvial-fan sequences owing to: (1) differences in interpretation regarding the response of alluvial-fan systems to particular climatic changes, (2) the relatively great sensitivity of alluvial-fan systems to environmental changes other than climate (Ritter, 1974; Bull, 1977), and (3) the small number of well-dated alluvial-fan sequences that can provide accurate time controls for regional correlations or for correlation with known chronologies of climatic change from the deep-sea record

(Shackelton and Opdyke, 1976). However, similarities in alluvial-fan sequences studied at several locations throughout the Basin and Range suggest that, in general terms, regional correlations and age assignments are possible. These similarities also suggest that the underlying factor controlling deposition was probably contemporaneous climatic change occurring over large regions.

Accordingly, in this paper we briefly discuss the criteria that are used to characterize alluvial-fan sequences and the potential problems with the use of individual criteria. We then present a three-fold classification for identifying, dating, and correlating alluvial-fan sequences in the Basin and Range. We emphasize the

TABLE 1. GENERAL CHARACTERISTICS OF YOUNG, INTERMEDIATE, AND OLD ALLUVIAL-FAN DEPOSITS

Characteristic	Young	Intermediate	Old
Drainage pattern	Distributary: anastomosing or braided	Tributary: dendritic	Tributary: dendritic or parallel
Depth of incision	Less than 1 m	Variable (1 to 10 m)	Greater than 10 m
Fan surface morphology	Bar-and-channel	Variable, generally smooth and flat	Ridge and valley, most of surface slopes
Preservation of fan surface	Presently active	Incised, but well preserved wide, flat divides	Basically destroyed, locally preserved on narrow divides
Desert pavement	None to weakly developed	None to strongly developed	None (surface destroyed) to strongly developed (surface preserved)
Desert varnish	None to weakly developed (most varnished clasts reworked from older surfaces or bedrock)	None to strongly developed	None (surface destroyed) to strongly developed (surface preserved)
B horizon	None to weakly developed	Weakly to strongly developed	None (surface destroyed) to strongly developed (surface preserved)
Calcic horizon	None to weakly developed, CaCO$_3$ disseminated throughout	Weakly to strongly developed	None, carbonate rubble on surface (surface destroyed) to strongly developed petrocalcic horizon (surface preserved)

Great Basin, Mojave Desert, and Sonoran Desert (Fig. 1), and we have not included the extensive work done in the Basin and Range of southern New Mexico.

ALLUVIAL-FAN UNITS AND RELATIVE-AGE CRITERIA

Alluvial-fan units defined in this study are based principally on relative-age criteria such as drainage and morphology; desert pavement, desert varnish, and soil-profile development; and morphostratigraphic relations with other Quaternary units. The most consistent and useful of these, particularly when mapping from aerial photographs, are drainage pattern, depth of stream incision, and fan surface morphology. Desert pavement, desert varnish, and soil-profile characteristics, particularly the stage of development of the B horizon and calcic horizon, are also useful, although they are subject to local variations from dependence on many factors other than age (Machette, this volume). Morphostratigraphic relations are important, particularly when ages of related units are known, and are time-equivalent on a regional scale. Examples of such units are deposits and shorelines of large pluvial lakes, tephra layers, and terraces of major through-flowing streams.

At least three general ages of alluvial-fan deposits are found in the Basin and Range province. These include young alluvial-fan deposits marked by active channels with depositional fan surfaces, intermediate-age deposits with well-preserved but inactive surfaces that are incised by modern channels, and old alluvial-fan deposits that have been highly dissected leaving virtually no trace of the original fan surface. Active channels may be mapped separately from active fan surfaces in most instances but for convenience are included with young deposits because they are an integral part of an active surface. Except for minor differences in grouping and terminology, these same three fan units have been used by Hamilton (1964), Bull (1974b), Shlemon and Purcell (1976), Shlemon (1978), Peterson (1981), and Hoover and others (1981). The basic characteristics of these units are shown in Table 1.

Drainage and Morphology

Conditions of erosion and deposition are important in making initial determinations of relative age. The predominance of either erosion or deposition is readily suggested by drainage pattern and morphology. For example, alluvial-fan deposits that are presently aggrading consist of a relatively closely spaced system of braided channels with an overall distributary drainage pattern. These characteristics typically describe the youngest alluvial-fan unit (Table 1; young alluvial-fan deposits). The depth of incision of drainages will generally be less than 1 m and will allow flooding of this young surface during major runoff events. These surfaces, particularly where gravelly, have flood-generated bar-and-channel topography (Fig. 2). In contrast, older alluvial-fan

Figure 2. Morphology and desert pavement characteristic of young (y), intermediate-age (i), and old (o) alluvial-fan deposits, northern Gila Mountains, southwestern Arizona.

deposits are undergoing degradation. Drainage on these older deposits is usually an incised dendritic or parallel tributary system. In these deposits, channel incision and fan surface morphology are useful in differentiating relative age. Channel incision depth generally increases with age. Correspondingly, the degree of preservation of original fan surfaces between incised channels generally decreases with age. Intermediate-age alluvial-fan deposits are thus incised but still retain much original fan surface topography (Figs. 2 and 3). In contrast, old alluvial-fan deposits have practically no original fan surface preserved and have interfluves that consist of subparallel ridges (Fig. 2).

Quantitative relations between age and morphologic characteristics are tentative because of the influence of other variables. For example, depth of incision depends on the size of the drainage basin and fan. Large fans with large drainage basins are generally more deeply incised than are smaller fans of roughly equivalent age from smaller drainage basins. Depth of incision and preservation of fan surfaces on interfluves are also influenced by local base-level effects. Alluvial-fan deposits close to an actively degrading base level have a greater depth of incision and poorer preservation of original fan surfaces than do fans of similar

age further from these base-level effects, or those in areas of stable or aggrading base levels. Tectonic base-level influences may also create morphologies which are not indicative of relative age (Bull, 1974b). Where fault scarps traverse alluvial deposits, the uplifted blocks between the fault and the mountain front are incised to depths that are more dependent on amount and rate of uplift than on age. Ranges of incision depths shown in Table 1 for each unit are given only as a general guideline, and depths outside the given ranges are not uncommon. Thus, although morphologic characteristics are useful to differentiate relative ages of deposits, other influences are also important for regional correlations.

Desert Pavement, Desert Varnish, and Soil-Profile Development

Several readily observable characteristics of desert soils are useful in evaluating the relative ages of alluvial-fan deposits. Probably the most diagnostic and easily recognized characteristics are desert pavement, desert varnish, and B-horizon and calcic-horizon development. Where soils occur on erosion surfaces cut on alluvial-fan deposits rather than on the original depositional

Figure 3. Surface morphology and stream incision on an intermediate-age alluvial-fan deposit, southern Snake Range, eastern Nevada.

surfaces, profile development only provides a minimum age of the deposits; in such cases, fan deposits may be much older than the soils suggest.

In deposits that contain gravel, a desert pavement may develop with time. Young alluvial-fan deposits in general lack pavements because depositional processes are still active. Old alluvial-fan deposits likewise lack pavements because the original fan surfaces have largely been removed by active erosional processes (Fig. 2). The presence of a smooth, well armored surface generally indicates an intermediate-age alluvial-fan surface. These surfaces have a wide range in pavement and varnish development, in part owing to grain-size and regional climatic variations and in part to the wide age range of the deposits. Pavements develop slowly in deposits that contain a small percentage of gravel, but develop very quickly in gravelly deposits and may, in fact, be found on young fans composed of coarse-grained alluvium. In the case of very gravelly deposits, the smoothness of the surface may be a better indication of relative age than is pavement development. The irregular bar-and-channel topography common in gravelly, young alluvial fans becomes progressively less pronounced with age once the surface is abandoned.

Intermediate-age fan surfaces, even in very coarse-grained alluvium, are generally quite smooth (Figs. 2 and 3).

With time, desert varnish will develop on exposed clasts in pavements. In addition to age, the amount of varnish is dependent on clast lithology, the availability of eolian dust, and the activity of manganese-concentrating organisms (Hunt, 1954; Engle and Sharp, 1958; Potter and Rossman, 1977, 1979; Perry and Adams, 1978; Dorn, 1983). A very dark varnish may develop on pavements composed of metamorphic and volcanic clasts, whereas varnish may be absent on pavements composed of granitic detritus, chiefly quartz and potassium feldspar. Color of the parent rock is also important because pavements composed of dark-colored clasts may appear more heavily varnished than they actually are. Climate and vegetation must also be considered, because on some fans in the high deserts of the Great Basin enough vegetation is present to preclude formation of pavement and varnish (Fig. 3). Thus, pavement and varnish development may be highly diagnostic of units in a particular area, but because of variations in grain size, clast lithology, climate and vegetation, and availability and composition of eolian dust, it may be difficult to apply these characteristics to regional correlations.

The accumulation of pedogenic calcium carbonate in the B and C horizons and the color and clay content of the B horizon in desert soils are further characteristics for determining relative ages of deposits. Gile and others (1966), Gile and Hawley (1972), Gile (1975, 1977), Bachman and Machette (1977), and Machette (this volume) discuss the progressive development of calcic horizons in the Basin and Range of New Mexico, and Bull (1974b) and Shlemon (1978) present detailed discussions of the development of desert soils in southeastern California and southwestern Arizona. Young alluvial-fan deposits lack soil-profile development because of their youth and aggrading conditions. Old fans may have a strong soil profile where original fan surfaces are preserved, but generally the soil profiles have either been entirely removed or partially truncated. In many instances, calcium carbonate clasts (caliche rubble) scattered on the present surface give evidence of the destruction of a petrocalcic horizon.

Intermediate-age alluvial-fan deposits have varying degrees of B-horizon (cambic to argillic) and calcic-horizon development (Bull, 1974b; Shlemon and Purcell, 1976; Shlemon, 1978; Hoover and others, 1981). Variations in thickness and degree of soil-profile development, particularly in the calcic horizon, stem from differences in lithology, availability of eolian material, climate, grain size, and other factors as well as age (Lattman, 1973; Bachman and Machette, 1977; Machette, this volume). Thus, although soil development can be highly diagnostic in helping differentiate alluvial-fan deposits, local variations in soil-profile development make regional correlations difficult.

Morphostratigraphic Relations

Geomorphic and stratigraphic relations between alluvial-fan deposits and other Quaternary sediments help determine relative age and in some cases help bracket absolute age. Alluvial-fan deposits in the Basin and Range are commonly associated with eolian sand, fluvial terraces, lacustrine deposits and shoreline features, and tephra and other volcanic deposits. The presence of stabilized eolian sand overlying an alluvial fan may indicate that the surface is no longer active. Furthermore, a geomorphic feature representing a local or regional base level to which an alluvial fan is or was graded may indicate relative age. For example, in open drainage basins, successively older fans may be graded to successively higher stream terraces. In closed, aggrading basins, bold surfaces commonly grade below present base level.

In many of these closed basins, morphostratigraphic relations between alluvial fans and pluvial lake shorelines and offshore deposits may indicate relative age. Where stream terraces or lake shorelines extend for great distances, they become useful in regional correlations. This is exemplified by stream terraces bordering the Gila and Colorado Rivers and the shorelines of Lake Bonneville and Lake Lahontan (Fig. 1). Tephra layers, such as the Bishop ash, which are interbedded with alluvial-fan deposits, also provide indications of alluvial-fan age (Shlemon and Purcell, 1976; Hoover and others, 1981).

AGE-DATING OF DEPOSITS

Several workers have successfully correlated alluvial-fan deposits based on relative-age criteria in the eastern Mojave and Sonoran Deserts of southeastern California and southwestern Arizona (Bull, 1974b; Lee and Bell, 1975; Shlemon and Purcell, 1976); western Mojave Desert of California (Ponti and others, 1980; Ponti, this volume); southern New Mexico (Gile, 1975, 1977); southern Nevada (Hoover and others, 1981); and the central Great Basin (Christenson and others, 1982). From these localities, dates have been obtained to calibrate relative-age criteria (Table 2).

Dates shown in Table 2 represent a wide variety of techniques used to date Quaternary deposits. In southeastern California and in southwestern Arizona, Shlemon and Purcell (1976) and Shlemon (1978) have dated deposits based on detailed descriptions of soil profiles and rates of soil-profile development calibrated from radiometrically and paleomagnetically dated sediments and from deposits of Bishop ash contained in the fan material. Bull (1974b) based his chronology on stratigraphic and geomorphic relations of fans with the Pliocene Bouse Formation and Colorado River terrace deposits, as well as on Th^{230}-U^{234} dating of pedogenic carbonates (Ku and others, 1979). Lee and Bell (1975) also relied on Th^{230}-U^{234} dates on pedogenic carbonate in alluvial-fan deposits and on geomorphic relations with Colorado River terraces containing bone fragments dated by amino-acid techniques.

In southern Nevada, Hoover and others (1981) used relations between alluvial-fan deposits, ash beds (including the Bishop ash), and Quaternary and older basalt flows in dating older fan deposits. They have also dated soils and pedogenic carbonates in younger units by the uranium-series and uranium-trend methods. The youngest units are dated by correlation with nearby radiocarbon-dated packrat middens and Holocene alluvial sequences. Similar methods were used by Sowers (1983) to date deposits of the Kyle Canyon alluvial fan in southern Nevada. A new technique utilizing paleomagnetism in pedogenic carbonate was applied in dating older deposits (Sowers, 1982).

In the central Great Basin, radiometrically dated Lake Lahontan and Lake Bonneville shorelines and offshore deposits have allowed dating of associated alluvial deposits. In the Lahontan basin, Hawley and Wilson (1965) indicate that their older alluvium predates Lake Lahontan Sehoo deposits and that the younger alluvium postdates these deposits. Morrison (1964, 1965) and Benson (1978) indicate the time of deposition of the Sehoo deposits to be between about 9,000 and 25,000 years ago. In the Bonneville basin, Christenson and others (1982) identify intermediate-age alluvial-fan deposits that predate the transgression of Lake Bonneville to the Bonneville shoreline, which had probably occurred by about 16,000 years ago (Currey, 1982). The young deposits postdate abandonment of the Bonneville shoreline about 14,000 years ago (Currey, 1982), and in many cases postdate the rapid recession of Lake Bonneville to lower levels as well.

TABLE 2. CORRELATION OF ALLUVIAL-FAN UNITS AND THEIR ESTIMATED AGES IN YEARS BEFORE PRESENT, BASIN AND RANGE PROVINCE

American Southwest Christenson and Purcell (this study)		Southeastern California and Southwestern Arizona					
		Eastern Mojave, California; Shelmon and Purcell (1976)		Eastern Mojave, California; Bull (1974b)		Colorado River, Arizona and California; Lee and Bell (1975)	
Unit	Age	Unit	Age	Unit	Age	Unit	Age
Young	0 to 10,000- 15,000	Q1 Q2	0 0-15,000	Q4 Q3	0 <11,000	Qal, Qf	<10,000
Inter- mediate	10,000-15,000 to 500,000- 700,000	Q3	15,000 to 500,000- 700,000	Q2	11,000- 200,000	Qfc	30,000- 100,000
Old	>500,000- 700,000	Q4	>500,000	Q1	500,000- >1,500,000	QTfc, QTfa	>500,000

Nevada Test Site; Hoover and others (1981) and Hoover (pers. commun., 1983)		Great Basin (Nevada and Utah)					
		Kyle Canyon, Southern Nevada; Sowers (1983 and pers. commun., 1983)		Lake Bonneville Basin[1]; Christenson and others (1982)		Lake Lahontan Basin[2], Winnemucca area; Hawley and Wilson (1965)	
Unit	Age	Unit	Age	Unit	Age	Unit	Age
Q1	< 8,000- 12,000	Surface 4	0-7,000	A5y	<14,000	Younger	<9,000
Q2	110,000- >730,000	Surface 3 Surface 2	7,000- 35,000 100,000- 300,000	A5i	>16,000	Older	>26,000
Qta	900,000- 1,100,000	Surface 1	>700,000				

[1]Lake Bonneville chronology from Currey (1982)
[2]Lake Lahontan chronology from Morrison (1964, 1965) and Benson (1978)

CONCLUSIONS

From studies throughout the Basin and Range, we group alluvial-fan deposits into young, intermediate, and old units and suggest that these are identifiable and correlative throughout the province. Dates shown in Table 2 indicate that young alluvial fans are generally Holocene or slightly older (less than 10,000 to 15,000 years), intermediate-age alluvial fans cover a broad range from late to middle Pleistocene (10,000 to 700,000 years), whereas old alluvial fans are older than 500,000 years and represent deposition during the early Pleistocene and late Tertiary.

The climatically controlled deposition of young (Holocene) alluvial fans has been discussed by Bull (1974a) and Wells (1977). However, where young alluvial-fan deposits have been dated on the basis of their occurrence below a particular pluvial lake level (Christenson and others, 1982) or on their occurrence stratigraphically above a particular lake deposit (Hawley and Wilson, 1965), they cannot be interpreted as direct evidence of climatically induced change in alluvial-fan deposition. The depo-

sition of these young alluvial-fan deposits is a consequence of lowered lacustrine base levels and is therefore only indirectly related to climatic change.

Regional climatic control over deposition of intermediate-age alluvial fans is less straightforward. These fans include a great variety of deposits which span a period of well-documented climatic change (late and middle Pleistocene time). Although these climatic changes resulted in several distinct levels of intermediate-age alluvial fans recognizable locally, our data are insufficient to attempt regional correlations of different fan levels. Old fans present a similar problem. Nevertheless, the general correlation in age of deposits with similar morphologies and soil development throughout the Basin and Range indicates that, within a certain range of variability, regional correlation and general age assignment (Holocene versus late to middle Pleistocene versus early Pleistocene) are possible. We hope that this tentative chronology will be further refined as more detailed alluvial-fan chronologies are worked out, particularly those dealing with intermediate-age fan deposits.

ACKNOWLEDGMENTS

The authors gratefully acknowledge the critical comments of Donald R. Currey, Roy J. Shlemon, and Genevieve Atwood, who reviewed the manuscript and offered helpful suggestions. This paper is in large part an outgrowth of work done while both authors were employed at Ertec Western, Inc. (formerly Fugro, Inc.), chiefly on projects related to the siting of nuclear power plants and MX missile facilities.

REFERENCES CITED

Bachman, G. O., and Machette, M. N., 1977, Calcic soils and calcretes in the southwestern United States: U.S. Geological Survey Open-File Report 77-794, 163 p.

Benson, L. V., 1978, Fluctuation in the level of pluvial Lake Lahontan during the last 40,000 years: Quaternary Research, v. 9, p. 300–318.

Bull, W. B., 1964, Geomorphology of segmented alluvial fans in western Fresno County, California: U.S. Geological Survey Professional Paper 352-E, p. 89–129.

—— 1974a, Effects of Holocene climate on arid fluvial systems, Whipple Mountains, California [abs.]: American Quaternary Association Abstracts, Third Biennial Meeting, p. 64.

—— 1974b, Geomorphic tectonic analysis of the Vidal region: Information concerning site characteristics, Vidal Nuclear Generating Station [California]: Los Angeles, Southern California Edison Company, Appendix 2.5B, amendment 1, 66 p.

—— 1977, The alluvial fan environment: Progress in Physical Geography, v. 1, p. 222–270.

Christenson, G. E., Miller, J. R., and Pieratti, D. D., 1982, Prediction of engineering properties and construction conditions from geomorphic mapping in regional siting studies, in Craig, R. G., and Craft, J. L., eds., Applied geomorphology, the "Binghamton" symposia in geomorphology—International Series, no. 11: London, George Allen and Unwin, p. 94–107.

Cooke, R. U., and Warren, A., 1973, Geomorphology in deserts: London, B. T. Batsford, Ltd., 374 p.

Currey, D. R., 1982, Lake Bonneville—selected features of relevance to neotectonic analysis: U.S. Geological Survey Open-File Report 82-1070, 31 p.

Denny, C. S., 1965, Alluvial fans in the Death Valley region, California and Nevada: U.S. Geological Survey Professional Paper 466, 62 p.

Dorn, R. I., 1983, Use of rock varnish as an arid lands geomorphology research tool, in Wells, S. G., and others, eds., Chaco Canyon Country: American Geomorphological Field Group, Guidebook, 1983 Conference, Northwestern New Mexico, p. 245.

Engle, C. G., and Sharp, R. P., 1958, Chemical data on desert varnish: Geological Society of America Bulletin, v. 69, p. 487–518.

Flint, R. F., 1971, Glacial and Quaternary geology: New York, John Wiley and Sons, 892 p.

Gile, L. H., 1975, Holocene soils and soil geomorphic relations in an arid region of southern New Mexico: Quaternary Research, v. 5, p. 321–360.

—— 1977, Holocene soils and soil-geomorphic relations in a semi-arid region of southern New Mexico: Quaternary Research, v. 7, p. 112–132.

Gile, L. H., and Hawley, J. W., 1972, The prediction of soil occurrence in certain desert regions of the southwestern United States: Soil Science Society of America Proceedings, v. 36, no. 1, p. 119–124.

Gile, L. H., Peterson, F. F., and Grossman, R. B., 1966, Morphological and genetic sequences of carbonate accumulation in desert soils: Soil Science, v. 101, p. 347–360.

Hamilton, W. B., 1964, Geologic map of the Big Maria Mountains Northeast quadrangle, Riverside County, California, and Yuma County, Arizona: U.S. Geological Survey Geologic Quadrangle Map GQ-350, scale, 1:24,000.

Hawley, J. W., and Wilson, W. E., 1965, Quaternary geology of the Winnemucca area, Nevada: University of Nevada, Desert Research Institute, Technical Report 5, 66 p.

Hoover, D. L., Swadley, W. C., and Gordon, A. J., 1981, Correlation characteristics and surficial deposits with a description of surficial stratigraphy in the Nevada Test Site region: U.S. Geological Survey Open-File Report 81-512, 27 p.

Hunt, C. B., 1954, Desert varnish: Science, v. 120, p. 183–184.

Ku, T. L., Bull, W. B., Freeman, S. T., and Knauss, K. C., 1979, Th^{230}-U^{234} dating of pedogenetic carbonates in gravelly soils of Vidal Valley, southeastern California: Geological Society of America Bulletin, Part I, v. 90, p. 1063–1073.

Lattman, L. H., 1973, Calcium carbonate cementation of alluvial fans in southern Nevada: Geological Society of America Bulletin, v. 84, p. 3013–3028.

Lee, G., and Bell, J., 1975, Depositional and geomorphic history of the lower Colorado River: San Diego Gas and Electric Company, Early Site Review Report, Sundesert Nuclear Power Project, Appendix 2.5-D, 19 p.

Machette, M. N., 1985, Calcic soils of the American Southwest, in Weide, D. L., ed., Soils and Quaternary geology of the southwestern United States: Geological Society of America Special Paper 203 (this volume).

Morrison, R. B., 1964, Lake Lahontan: Geology of the southern Carson Desert, Nevada: U.S. Geological Survey Professional Paper 401, 156 p.

—— 1965, Quaternary geology of the Great Basin, in Wright, H. E., Jr., and Frye, D. G., eds., The Quaternary of the United States: Princeton, Princeton University Press, p. 265–286.

Perry, R. S., and Adams, J. B., 1978, Desert varnish—evidence for cyclic deposition of manganese: Nature, v. 276, p. 489–491.

Peterson, F. F., 1981, Landforms of the Basin and Range province defined for soil survey: Nevada Agricultural Experiment Station Technical Bulletin 28, 52 p.

Ponti, D. J., 1985, The Quaternary alluvial sequence of the Antelope Valley, California, in Weide, D. L., ed., Soils and Quaternary geology of the southwestern United States: Geological Society of America Special Paper 203 (this volume).

Ponti, D. J., Burke, D. B., Marchand, D. E., Atwater, B. F., and Helley, E. J., 1980, Evidence for correlation and climate control of sequences of late Quaternary alluvium in California [abs.]: Geological Society of America Abstracts with Programs, v. 12, p. 501.

Potter, R. M., and Rossman, G. R., 1977, Desert varnish—the importance of clay minerals: Science, v. 196, p. 1446–1448.

—— 1979, The manganese and iron oxide mineralogy of desert varnish: Chemical Geology, v. 25, p. 74–79.

Ritter, D. F., 1974, Holocene climate change and fluvial systems [abs.]: American Quaternary Association Abstracts, Third Biennial Meeting, p. 58–62.

Shackelton, N. J., and Opdyke, N. D., 1976, Oxygen-isotope and paleomagnetic stratigraphy of Pacific core V28-239, late Pliocene to latest Pleistocene, in Cline, R. M., and Hayes, J. D., eds., Investigation of late Quaternary paleoceanography and paleoclimatology: Geological Society of America Memoir 145, p. 449–464.

Shlemon, R. J., 1978, Quaternary soil-geomorphic relationships, southeastern Mojave Desert, California and Arizona, in Mahaney, W. C., ed., Quaternary Soils: Norwich, University of East Anglia, Geo Abstracts Ltd., p. 187–207.

Shlemon, R. J., and Purcell, C. W., 1976, Geomorphic reconnaissance, southeastern Mojave Desert, California and Arizona: San Diego Gas and Electric Company, Early Site Review Report, Sundesert Nuclear Power Project, Appendix 2.5-M, 28 p.

Sowers, J. M., 1982, Remnant magnetism of pedogenic calcretes from southern Nevada [abs.]: EOS (Transactions of the American Geophysical Union), v. 63, p. 918.

—— 1983, The Quaternary history of the Kyle Canyon alluvial fan, southern Nevada, in Wells, S. G., and others, eds., Chaco Canyon Country: American Geomorphological Field Group, Guidebook, 1983 Conference, Northwestern New Mexico, p. 250–251.

Wells, S. G., 1977, Geomorphic controls of alluvial fan deposition in the Sonoran Desert, southwestern Arizona, in Doering, D. O., ed., Geomorphology in arid regions: Eighth Annual Geomorphology Symposium, State University of New York, Binghamton, New York, p. 27–50.

Manuscript Accepted by the Society January 12, 1985

Geological Society of America
Special Paper 203
1985

Pliocene/Quaternary geology, geomorphology, and tectonics of Arizona

Roger B. Morrison
Morrison and Associates
13150 West 9th Avenue
Golden, Colorado 80401

ABSTRACT

This paper summarizes four decades of my research on the late Cenozoic geology of Arizona that culminated in several maps of late Pliocene and Quaternary geology and neotectonic features of the whole state. Principal conclusions are: (1) Styles of deposition, preservation, and exposure of Pliocene to Holocene deposits differ greatly in various parts of the state. The Colorado Plateau has vast areas of stripped bedrock with local patches of late Quaternary eolian sand and alluvium and also large areas of Bidahochi Formation (Miocene and Pliocene, alluvial-lacustral), as well as the San Francisco, Toroweap, and Springerville-Showlow volcanic fields. South of the Mogollon Rim, a zone 50 to 100 km wide is deeply dissected terrain with a few tiny patches of Quaternary deposits. The main Mexican Highland section of the Basin and Range province contains the fullest, best-exposed Pliocene through Quaternary sequences in the state, in numerous well-dissected intermontane basins, as well as the San Bernardino basalt field. The Sonoran Desert section has widespread alluvium and local eolian sand in broad, generally little-dissected intermontane basins, and also contains several basalt fields. (2) Quaternary sedimentation shows a primary cyclicity of several hundred thousand years, a dominant rhythm slower than marine oxygen-isotopic glacial cycles, with mainly climatic, not tectonic, control. However, except for the Holocene, many details of depositional chronology still are poorly understood and poorly correlated. (3) We found approximately 144 proven and probable faults less than 3.5 m.y. old, chiefly in a 200-km zone from the northwest to southeast corners of the state. Surface displacements are mostly older than 100,000 years, and we found none younger than 4,000 years. Few faults show multiple displacements. Recurrence intervals are 10^5 years for most faults, to 10^4 years in some cases; however, parts of northwestern, southwestern, southeastern, and central Arizona have regional recurrence intervals of 10^3 years.

INTRODUCTION

This paper summarizes four decades of my observations on the late Cenozoic geology of Arizona, culminating with two projects from 1978 to 1982 that focused on mapping upper Pliocene and Quaternary neotectonic features and lithologic and stratigraphic units throughout the state (Morrison and others, 1981a, 1981b; Morrison and Menges, 1982). In these latter projects, I was ably assisted by Chris Menges, then of the Arizona Bureau of Geology and Mineral Technology. In preparing this summary and the maps on which it is based, I have tried to evaluate and use

the products of other geologists. This commonly requires compromise, a situation best defined by Ambrose Bierce (1911): "Such an adjustment of conflicting interests as gives each adversary the satisfaction of thinking he has got what he ought not have, and he is deprived of nothing except what was justly his due." Nevertheless, our mapping is mostly original work, because few previous geologic maps show the Pliocene and Quaternary lithologic and stratigraphic units and neotectonic features in adequate detail and with reliable age discrimination. Our mapping is based on interpretation of U-2 high-altitude air photos (scale about 1:125,000), previous field experience in Arizona, extensive ground reconnaissance, and low-altitude aircraft reconnaissance flights. The systematic, state-wide photointerpretive mapping, closely controlled by ground observations, provided a synoptic overview of the late Cenozoic surficial geology and geomorphology of Arizona.

As chronologic benchmarks for the following discussions of late Cenozoic events, I have used available radiometric dates, geomagnetic polarity data, and mammalian paleontologic data. Because these data still are meager, although much increased during the past two decades, I have been obliged to supplement the numerical controls extensively with my own extrapolations. I recognize that my subjective input may lead to uncertainties and differences of opinion, and I hope that the controversial items will stimulate further research.

Salient features of the surficial (Pliocene through Quaternary) geology and geomorphology of Arizona are:

(1) Styles of landscape development and assemblages of surficial deposits differ significantly between Arizona's major physiographic regions: The Colorado Plateau and the three parts of the Basin and Range province—the main Mexican Highland section, the transition zone between the Mexican Highland and the Colorado Plateau, and the Sonoran Desert section (Fig. 1). Variation in style and history of later Cenozoic tectonism in each region has caused pronounced differences in relief, climate, patterns of erosion and deposition, and chronology of landscape development.

Figure 1. Map of Arizona showing its principal physiographic regions.

(2) Arizona landscapes commonly show much older land surfaces and associated surficial deposits and soils (and, consequently, much greater diversity in age of these features) than is usual for more northerly and more humid regions, especially glaciated and periglacial ones.

(3) Alluvium generally dominates, although eolian sand locally is abundant on the Colorado Plateau and in southwestern Arizona. Likewise, basaltic flows, cinders, and lapilli commonly predominate in eight widely scattered volcanic fields. Colluvial deposits (except landslide debris) are too sparse to be shown at our scales of mapping. Lacustrine deposits are restricted to the interiors of some intermontane basins and small parts of the Colorado Plateau. Glacial till and outwash are almost infinitesimal patches in the two highest mountain areas.

SURFICIAL GEOLOGY AND GEOMORPHOLOGY OF THE COLORADO PLATEAU (ARIZONA PORTION)

The Colorado Plateau in Arizona comprises a series of extensive plateaus (Mogollon, Coconino, Shivwits, Uinkaret, Kanab, Kaibab, Paria, Kaibito, Moenkopi, and Defiance Plateaus, and Black Mesa) 1,500 to 2,750 m in altitude. It also includes the White Mountains-Springerville-Showlow, San Francisco, and Uinkaret volcanic fields, the first two with peaks above 3,000 and 3,650 m, respectively. In addition to its considerable elevation, nearly all above 1,500 m, distinctive features of the Colorado Plateau include: (1) Many deep canyons, more than in any other part of the United States. Chief among these is the Grand Canyon of the Colorado River. (2) Arid to semiarid climate, with mean annual precipitation mostly between 150 and 400 mm and mean annual temperature of 8° to 12°C, ranging to 750 mm and about 0°C in the higher mountains. Spasmodic torrential rains and sparse vegetation foster maximum runoff, favoring sheetwash erosion and local arroyo trenching. (3) Approximate horizontality of the bedrock. Strongly inclined beds are limited to the few great monoclines and the borders of certain uplifts. Exposed bedrock is chiefly upper Paleozoic limestones and sandstones, Mesozoic sandstones and shales, and Neogene basaltic flows and pyroclastics, soft sandstones, and siltstones. (4) Strong steplike angularity of landscape displayed over great distances: innumerable cliffy escarpments separated by wide gentle slopes. This is the product of differential erosion of the generally flat-lying rocks under the arid climate—steep retreating scarps capped by resistant rocks, with structural benches atop the escarpments that grade upslope onto extensive pediments developed on weak rocks. Also characteristic are wide plains, almost devoid of surficial deposits, underlain by resistant rocks. (5) Apart from the escarpments, the great relief is due chiefly to incision of deep canyons below moderately flat terrain (Figs. 2, 3, and 4). The three volcanic fields are superposed on the sedimentary rocks of individual plateaus, and the high mountains in the White and San Francisco Mountains are purely local volcanic piles.

Figure 2. Monument Valley. Cliffy escarpments in thick-bedded sandstones rise above pediments cut across shaly units (Permian and Triassic). The pediments have extensive mantles of eolian sand. Slender spires are the result of several million years of back-wasting erosion. Obviously they could not have survived any large earthquakes. They, therefore, document the aseismicity of this region for much of Quaternary time.

Figure 3. The Grand Canyon of the Colorado River, viewed from its southern rim (north edge of the Coconino Plateau). The Kaibab Plateau (at left skyline) is 2,400 to 2,850 m in altitude, as much as 2,000 m above the Colorado River (seen left of center of view).

Late Cenozoic Erosional History

In this huge region, surficial deposits are surprisingly scanty; the bedrock is stripped of mappable surface cover over vast areas. This is the result of the whole Colorado Plateau being structurally and topographically high through the middle and late Cenozoic due to epeirogenic uplift (associated with the East Pacific Rise) that started in the middle Eocene and lasted into later Miocene (Damon, 1971; Shafiqullah and others, 1978, 1980; P. E. Damon, written commun., 1984).

Various erosion cycles punctuated the Cenozoic erosional history. This history has been most studied in the part of the Colorado Plateau east and northeast of the San Francisco Mountains (Gregory, 1917; McCann, 1938; Childs, 1948; Cooley and Akers, 1961a; Cooley and others, 1969; Damon and others, 1974). Here, five Miocene to Quaternary erosion cycles are recognized. They are named after scattered remnants of erosion surfaces that formed during the chiefly lateral-planation (pedimentation) phases of each cycle: the Valencia (oldest), Hopi Buttes, Zuni, Black Point, and Wupatki surfaces. The older erosion surfaces rarely bear alluvium commensurate with their ages, although in places they bear younger eolian, or alluvial and eolian deposits. In some instances, the erosion surfaces are preserved beneath lava flows.

The Valencia surface developed prior to 9.40±0.91 m.y. ago (Damon and others, 1974, p. 226), likely in the middle Miocene (Cooley and Akers, 1961b; Cooley and others, 1969), but perhaps starting somewhat earlier. Its relief was low, and principal streams were superposed across the earlier bedrock structures. The Valencia and earlier Tertiary erosional episodes comprise the "Great Denudation" of Dutton (1882) and the "Plateau Cycle" of Davis (1901). During the Valencia cycle, the courses of the present principal streams were established approximately, although this cycle predates the cutting of the Grand Canyon.

During the early part of the Hopi Buttes cycle, in the late Miocene, accelerated downcutting entrenched the ancestral Colorado and Little Colorado River systems 300 to 450 m below the Valencia surface. Then a huge lake, Lake Bidahochi, was gradually created in the broad ancestral valley of the Little Colorado River, perhaps because of rejuvenation of the Kaibab uplift or because of damming of major streams by early volcanism in the San Francisco volcanic field. Fluvial downcutting lessened

Figure 4. Canyon de Chelly. This canyon is incised into a gently west-sloping plateau on resistant Permian and Triassic sandstones that is nearly free of surficial deposits.

markedly, lateral planation increased, and the Hopi Buttes erosion surface was formed in the uplands above where lacustrine sediments of the lower member of the Bidahochi Formation accumulated in Clarendonian time.

A volcanic episode in the Hopi Buttes area, mainly 8.5 to 4.2 m.y. ago (P. E. Damon, personal commun., 1983), produced more than 200 diatremes and associated flows and pyroclastics of feldspathic basalt to carbonatite and coeval lake sediments, comprising the middle member of the Bidahochi Formation. Both boundaries (upper and lower) of this member appear to be time transgressive with the upper and lower members of the Bidahochi Formation.

The relatively minor Zuni erosion cycle, early Pliocene, is represented by alluvial deposits on the Zuni erosion surface and by the essentially coeval upper member of the Bidahochi Formation, which consists of alluvial, some lacustrine, and minor eolian sediments of Hemphillian age, deposited in the broad valley of the Little Colorado River.

Lake Bidahochi disappeared about 4 m.y. ago, probably by becoming completely drained by entrenchment of the Colorado River system, particularly after the Grand Canyon eroded head-

ward across the Kaibab uplift. There is no evidence that Lake Bidahochi, during its approximate 5 m.y. of life, ever drained eastward across New Mexico (J. W. Hawley, personal commun., 1983).

Accelerated downcutting initiated the Black Point erosion cycle and the beginning of the major excavation of the principal valleys in northeastern Arizona. This erosion cycle lasted from later Pliocene (Blancan) into early Pleistocene (Irvingtonian) time. It had two main oscillations which produced, during temporary reduction in rate of downcutting in the latter parts of each oscillation, two main pediment-terrace surfaces that locally bear alluvial gravel. This erosion cycle is named after the Black Point basalt flow, K-Ar dated at 2.39±0.32 m.y. ago (Damon and others, 1967, 1974), which flowed onto the early Black Point erosion surface, damming the Little Colorado River. Since this eruption, the Little Colorado River has downcut about 200 m, an erosion rate that I estimate to be about 84 m per m.y. [My estimate is slightly lower than the 90 m per m.y. proposed by Damon and others (1974). On the other hand, I estimate that the Little Colorado River has cut 66 to 79 m below the top of the 0.51-m.y.-old Tappan Spring Canyon flow that dammed the river

Figure 5. Low escarpments and pediments developed in mainly mudstone/shale units (Triassic; near Round Rock, Chinle Valley). The pediments locally have thin veneers of eolian sand, colluvium, and alluvium, usually less than 1 m thick.

2.4 to 9.6 km below Cameron. This suggests an erosion rate of 142±13 m per m.y., which is slightly higher than the estimates of Damon and others (1974) and Cooley and Wilson (1968) for this site.]

The final erosion cycle, the Wupatki cycle, has continued to the present. It is characterized by regional downcutting, interrupted, due to climatic change, by five main episodes of lessened downcutting, pedimentation, and stream-terrace formation. One episode occurred in the late early Pleistocene, two in the middle Pleistocene, and two in the later Pleistocene.

The late Cenozoic erosional history of the western part of the Colorado Plateau is somewhat more complex but has been little studied. It was influenced by faulting and local volcanism and, starting with the Pliocene, by development of the Colorado River and the Grand Canyon. The Colorado River at the mouth of the Grand Canyon (at the Grand Wash Cliffs, the western edge of the Colorado Plateau) is less than 5.9 m.y. old, but by 3.8 m.y. ago it had cut within 110 m of its present depth here (Damon and others, 1978), and by 1.16 m.y. ago it had cut to its present depth at Lava Falls (McKee and others, 1968).

Despite this long erosional history, mappable alluvial depos-

its are few. The gentler valley slopes below the escarpments are pediments cut on bedrock, mostly on the weaker shaly units (Fig. 5). These slopes rarely have mappable alluvial or colluvial mantle, for bedrock is typically within 1 m of the land surface. True rock pediments thus are much more ubiquitous and characteristic of the Colorado Plateau than of the Sonoran Desert. The main alluvial areas are the flood plains and low terraces (Holocene and late Pleistocene) along the larger streams (except the Colorado River) and usually are narrow, discontinuous, very crenulate, and usually less than 15 m thick. Isolated remnants of middle Pleistocene terrace gravel occur along some reaches of the Little Colorado and, rarely, the Colorado Rivers. Still older alluvial gravels and accumulations of lag gravel occur in widely scattered remnants—significant indications of ancestral drainage systems, for example, those on the Black Point erosion surface.

Eolian Deposits

Eolian deposits are common in a 34,000 km^2 area comprising much of the Navajo-Hopi country in northeastern Arizona. The western boundary of this area runs along the Little

Colorado River north and northeast from St. Johns to Cameron and thence along the Echo Cliffs north past Page. Excluded is Black Mesa and a 16- to 48-km-wide belt of uplands west of the New Mexico border and east of a line from the southern boundary of the Navajo Reservation to Ganado, Chinle, and thence to Four Corners. Nearly all the dunes are seif, trending N45E to N65E. Airphotos commonly show a strong linear pattern of parallel sand streaks, with some individual dune forms extending many kilometers over irregular bedrock surfaces. However, a belt of barchan and transverse dunes lies along the western edge of this sand-rich area, extending from about 65 km south to 40 km north of Tuba City. Eolian deposits are essentially absent on the central and western parts of the Colorado Plateau in Arizona. The eolian deposits are chiefly eolian sand, but in places include variable amounts of colluvially and fluvially reworked sand and silt. The exposed deposits are mostly Holocene and late Pleistocene, but local exposures of strongly developed buried soils show that some deposits are middle Pleistocene.

Glacial Deposits

The only glacial deposits in Arizona, totaling about 18 km^2 in area, are in the San Francisco Mountains (Sharp, 1942; Péwé and Updike, 1976) and White Mountains (Merrill and Péwé, 1977). In the San Francisco Mountains, glaciers extended down four main radial valleys, as much as 7 km from the cirque headwalls. In the White Mountains (Mts. Ord and Baldy), glaciers extended down five main radial valleys a maximum of 7.5 km. Three ages of Pleistocene till were recognized in each area and correlated respectively with the Sacagawea Ridge-Mono Basin (the outermost end moraines), the Bull Lake-Tahoe, and the Pinedale—Tenaya-Tioga-Hilgard Glaciations in the Rocky Mountains and Sierra Nevada. This "finger-matching" correlation may be questioned because outwash of the oldest glaciation in the San Francisco Mountains postdates emplacement of a rhyolite volcano 210,000 years ago (Updike and Péwé, 1974; Péwé and Updike, 1976; Damon and others, 1974); thus, this glaciation could be correlated to the Bull Lake-Tahoe, at least in part. Tiny Holocene moraines also occur in the White Mountains, as well as Holocene periglacial deposits in the San Francisco Mountains.

Volcanic Fields

Three volcanic fields on the Colorado Plateau contain Pliocene and Quaternary lavas and pyroclastics. The Springerville-Showlow volcanic field, about 5,000 km^2, mainly north and northeast of the White Mountains, still is little studied. It has about 200 cinder and strato-volcanic cones ranging in age from Pliocene to late Pleistocene. It also includes extensive basalt flows. The older ones, 6.03 to 1.76 m.y. old (Laughlin and others, 1979, 1980; Luedke and Smith, 1978; Peirce and others, 1979), came from the White Mountains volcanic field to the south (whose main activity was in the later Miocene); most flowed

northward but several of middle to terminal Pliocene age flowed southwestward past the Mogollon Rim, down Corduroy Creek and White River Canyon (Peirce and others, 1979). Younger flows, 1.67 to 0.75 m.y. old (Laughlin and others, 1979, 1980), appear to be more limited and flowed either northward or radially from local vents. One small flow, 13 km northwest of Springerville, appears to be of late Pleistocene age. Both tholeiitic and alkalic basalts were erupted, and there is no correlation between age and composition.

The San Francisco volcanic field covers about 7,800 km^2 centered around the San Francisco Mountains. The Pliocene and Quaternary volcanics are mainly basalt flows, cinders, and lapilli (with several hundred cinder cones), but also include several strato-volcanoes (the San Francisco Mountains complex, O'Leary and Kendricks Peak, and Sitgreaves Mountain), in places with small areas of andesitic flows and pyroclastics, and rhyodacite, latite, and rhyolite flows, tuff, and plugs. The younger activity, middle Pleistocene to Holocene, is all in the eastern half of the field (Damon and others, 1974; Luedke and Smith, 1978). The youngest eruption, at Sunset Crater, occurred about 1066 A.D. (Smiley, 1958). Especially characteristic of this volcanic field is the large number of depressions, commonly poorly drained, caused by disruption of drainage by lava flows and faults.

The Uinkaret volcanic field, in the Uinkaret/Toroweap Valley north of the Colorado River, has both feldspathic undersaturated and normal basaltic flows and pyroclastics, erupted during Pliocene through late Pleistocene time from north-trending swarms of vents. At least 150 cinder cones are present. One of the largest, Vulcan's Throne, erupted about 0.1 m.y. ago (only an order-of-magnitude age estimate; Damon and others, 1967). Numerous lava flows cascaded into Grand Canyon (temporarily damming the Colorado River), most recently about 0.1 to 0.2 m.y. ago (Hamblin, 1974).

SURFICIAL GEOLOGY AND GEOMORPHOLOGY OF THE BASIN AND RANGE PROVINCE IN ARIZONA

General Background

The Basin and Range province contrasts strikingly with the Colorado Plateau in its physiography, surficial geology, and geomorphology. Its principal characteristic is a series of discontinuous mountain ranges alternating with subparallel intermontane basins, trending west-northwest to north and, subordinately, north to northeast. The higher summits in the mountains vary from 1,600 to 3,265 m in altitude in the eastern part of the state, and from 900 to 2,520 m near the Colorado River. The intermontane basin floors are much lower than the Colorado Plateau and generally much more arid. This province has two main divisions in Arizona, the Mexican Highland section and the Sonoran Desert section. In both divisions, the mountain ranges display great diversity in structure, age, and lithology of their rocks, even between nearby ranges. Exposed rocks range from Precambrian to Quaternary and include virtually all lithologies: sedimentary, volcanic, plutonic, and metamorphic. Some ranges are structurally and lithologically simple, whereas other mountain ranges have complex internal structure, commonly involving several episodes of low-angle faulting, high-angle faulting, folding, and (or) igneous intrusion—and similarly complex stratigraphy and lithology.

The Tertiary history of this region, following the final stage of the Laramide orogeny, can be divided into four fairly well defined episodes (Scarborough and Peirce, 1978): (1) Tectonic and magmatic quiescence persisted during most of the Eocene and Oligocene, with development of a very widespread low-relief erosion surface. (2) The interval from about 32 to 20 m.y. ago (the mid-Tertiary orogeny of Damon and Bickerman, 1964; and Shafiqullah and others, 1976, 1980) was characterized by major calcalkaline volcanism, plutonism, and associated sedimentation. (3) Between about 20 and 12±1 m.y. ago, minor post-orogenic volcanism, complicated tectonism, and important sedimentation took place. (4) About 12±1 m.y. ago, a different style of tectonism began, the Basin and Range deformation *sensu strictu* (the Basin and Range disturbance of Scarborough and Peirce, 1978; Anderson and others, 1972; Shafiqullah and others, 1980; Eberly and Stanley, 1978; Menges and McFadden, 1981; H. W. Peirce, written commun., 1982).

The Basin and Range deformation produced the larger elements of the present topography in the Basin and Range portion of Arizona (Fig. 6). The transition into this deformation episode was marked by cessation of listric faulting and andesitic volcanism, and a change to high-angle normal "block" faulting. Also, orientation of active faults (mostly reactivated earlier faults) changed to dominantly northerly, although ranging from west-northwest to north to northeast. Horsts and grabens were produced with a wide range of displacements and lateral dimensions. The overall result is that the principal basin-margin fault zones commonly have a zig-zag pattern, with some segments trending northeast or northwest, in contrast to the general north to north-northwest trend. The resulting intermontane structural basins contain later Miocene syntectonic sediments that grade upward into relatively undeformed Pliocene deposits. These basin-fill sequences locally attain maximum thickness of more than 3,000 m.

The Basin and Range deformation ended earlier in the Sonoran Desert (10.5 to 6 m.y. ago) than in the Mexican Highland section (about 6 to >3 m.y. ago) (Shafiqullah and others, 1980; Eberly and Stanley, 1978; Menges and McFadden, 1981; H. W. Peirce, written commun., 1982). In both sections, however, high-angle normal faulting continued on a much-reduced, gradually diminishing, and more localized scale. In the Sonoran Desert, significant faulting ceased early in the Pliocene, resulting in the oldest landscapes in the entire Basin and Range province. In the Mexican Highland section and along the western edge of the Colorado Plateau, however, infrequent faulting continued in places into the late Pleistocene and perhaps into early Holocene.

Surficial Geology and Geomorphology of the Transition Zone of the Mexican Highland Section

A belt 65 to 80 km wide, south of the Mogollon Rim (the southern boundary of the Colorado Plateau) and extending from the Tonto Creek–Lake Roosevelt basin southeastward to the New Mexico line, is here termed the transition zone of the Mexican Highland section of the Basin and Range province, because it contains features common to both the Colorado Plateau and the Basin and Range province (Fig. 7). It is mostly a maze of deep canyons and high narrow ridges. In places, however, there are remnants of low-relief erosion surfaces (pediments, locally mantled with coarse alluvial gravel) that are commonly hundreds of feet above principal streams. These surfaces probably are Pliocene and correlate with the Sonoita and Martinez surfaces, discussed below. The high basin-like areas that contain these surfaces appear to have been depressed by high-angle faults that were activated early during the Basin and Range deformation and then became relatively quiescent.

Quaternary deposits are sparse in this rugged area. They consist of narrow, very discontinuous patches of Holocene alluvium in the wider parts of a few canyons, and miniscule remnants of upper and middle Pleistocene terrace gravel in even fewer canyons.

Surficial Geology and Geomorphology of the Main Mexican Highland Section

General Features. The main part of the Mexican Highland section is characterized by generally north-northwest to north-trending mountain ranges, 45 to 95 km long, and intervening intermontane basins which are longer than the individual bordering mountain ranges. The basins and mountain ranges are roughly equal in width, each class ranging between 8 and 30 km wide. The range fronts are generally straight or gently curving in plan, with moderately embayed valleys. The mountain ranges have many summits above 1,500 m, and the higher ones are over 2,700 m in altitude; the highest, Mt. Graham, in the Pinaleno Range, is 3,265 m high. Basin floors along axial streams are mostly between 600 and 1,370 m in altitude; a few bordering the Sonoran Desert extend down to 450 m. All the basins have exterior drainage except two: Hualapai Valley (with Red Lake Playa) and central Sulfur Spring Valley (with Wilcox Playa).

Pliocene Geomorphic History, Geomorphic Surfaces, and Deposits. The intermontane basins of the main Mexican Highland section have the deepest, most fully exposed Pliocene and Quaternary sequences in the state. These basins apparently had interior drainage during most or all of the later Miocene episode of Basin and Range deformation *sensu strictu*. During this deformation, large quantities of alluvium were dumped rapidly into the basins, forming thick sequences of gravel and sand of monotonous lithology, grading to silt, clay, marl, and other paludal and lacustrine sediments in the lower parts of the basins. Up-section, these deposits are progressively less faulted and tilted

Figure 6. Landsat-1 multispectral image (ERTS-E 1193-17330, band 5, taken 1 Feb., 1973) showing parts of the Colorado Plateau and the transition zone of the Mexican Highland section of the Basin and Range province (dotted line indicates their boundary at the Mogollon Rim). F = Flagstaff, P = Payson, W = Winslow, H = Holbrook, I-40 = Interstate Highway 40, LC = Little Colorado River, MC = Meteor Crater, MH = Mt. Humphreys (highest peak in the San Francisco Mountains), ML = graben of Upper and Lower Marys Lakes, and HB = Hopi Buttes area.

from their original depositional dips, less strongly jointed, and less cemented, and their lithologic makeup is increasingly in accord with the local source areas.

Near the start of the Pliocene (5.6 m.y. ago, as recently dated), the Basin and Range deformation died down, and climate began to predominate over tectonism in controlling erosion and sedimentation. This was a time of generally equable warm-semiarid climate in this region, without the large-amplitude climatic cycles that characterized the Quaternary. As a result, erosion-deposition rates decreased by an order of magnitude, the alluvium became generally finer, and dozens of small diastems, commonly marked with paleosols, formed intermittently during times of landscape erosional stability (see *Gardner Canyon Alloformation* below). The fluvial systems fluctuated close to their

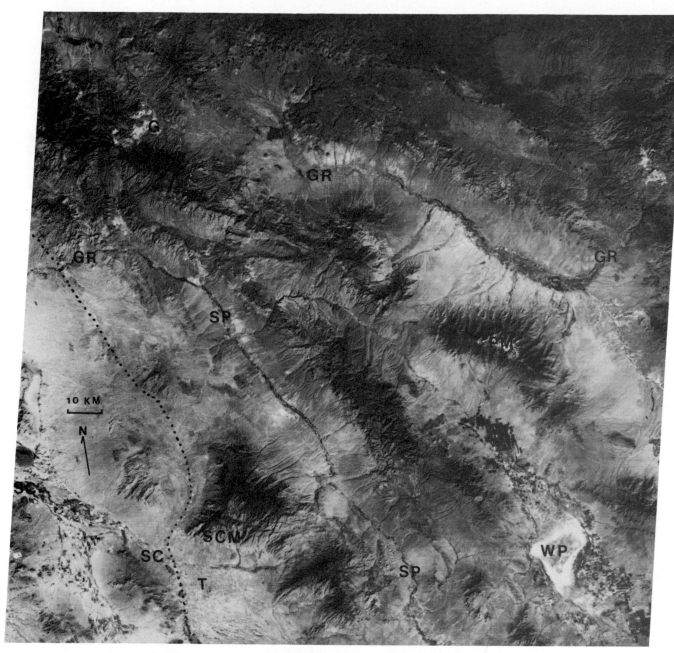

Figure 7. Landsat-1 multispectral image (ERTS-E 1102-17274, band 5, taken 2 Nov., 1972) showing parts of the main Mexican Highland section and small parts of the transition zone of this section (in the northeastern corner) and of the Sonoran Desert section (at southwestern corner of the image). Dotted lines indicate the boundaries between these physiographic units. T = Tucson, G = Globe, S = Safford, WP = Wilcox Playa, GR = Gila River, SP = San Pedro River, SC = Santa Cruz River, SCM = Santa Catalina Mountains, and PM = Pinaleno Mountains.

critical-power threshold (Bull, 1979) for about 4 m.y. (Menges and McFadden, 1981), until soon after the onset of the Pleistocene. This combination of tectonic and climatic tranquility resulted in the chief episode of pedimentation throughout the Mexican Highland section during the entire late Cenozoic. It produced pediments commonly 2 to 5 km wide, even on resistant gneissic, granitic, volcanic, and sedimentary rocks.

This pediment surface is widely (albeit fragmentarily) preserved in the Mexican Highland section. It is exposed locally at many mountain margins as small erosional remnants, usually considerably dissected. It also is preserved in the subsurface of certain parts of intermontane basins where the younger cover has not been eroded. Its remnants in the Santa Catalina-Tortolita Mountains area have been called the "Rillito surface" (Pashley,

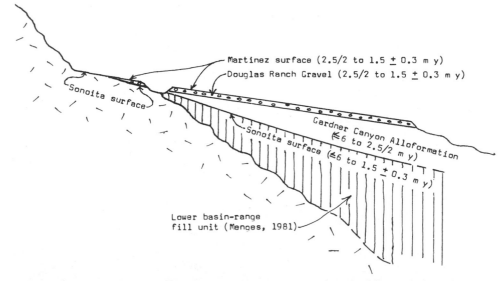

Figure 8. Diagram showing the relations of the four geomorphic/stratigraphic units that document the Pliocene erosional/depositional history in the Mexican Highland section. These are the Sonoita and Martinez surfaces (pediment-terrace geomorphic surfaces) and the Gardner Canyon and Douglas Ranch Alloformations.

1966; Budden, 1975). However, the name "Rillito beds (I, II, and III)" also has been used for three sedimentary units in this area, ranging in age from Oligocene to Miocene(?) (Pashley, 1966). This important regional geomorphic surface deserves an unambiguous name. Therefore, I propose that it be called the *Sonoita surface.* This name is appropriate because the stratigraphic/chronologic relations of this surface to underlying and overlying sediments and soils have been studied in most detail in the Sonoita area, where they are exceptionally well displayed (Menges, 1981; Menges and McFadden, 1981). In this area, the Sonoita surface began to form at the same time as the start of deposition of the "upper basin-range fill" and continued forming while the Martinez surface was developing (Menges, 1981), until about the beginning of the Pleistocene—a total span of about 4 m.y. Various stratigraphic, geomorphic, and chronologic data pertaining to this important prelude to Quaternary landscape history need clarification and therefore will be discussed below.

Four main geomorphic-stratigraphic entities record the Pliocene erosional-depositional history of the Mexican Highland section (Figs. 8 and 9): (1) The *Sonoita surface,* mentioned above (this geomorphic surface is the most time-transgressive of the four units). (2) A sedimentary unit that overlies much of the Sonoita surface, which I propose to call the *Gardner Canyon Alloformation.* [The term "alloformation" is the fundamental unit in a new category of stratigraphic classification (North American Commission on Stratigraphic Nomenclature, 1983). An allostratigraphic unit is a mappable stratiform body of sedimentary rock that is defined and identified on the basis of its bounding discontinuities, rather than by content. Its boundaries are laterally traceable discontinuities.] The Gardner Canyon Alloformation, in its type area, the upper Sonoita Creek basin, is the "upper basin-range fill" of Menges (1981), which is the same as the "upper basin fill"

of Menges and McFadden (1981). It overlies the "lower basin-range fill" of Menges (1981); it is generally less than 100 m thick and onlaps onto the Sonoita surface. Its age as determined by magnetostratigraphy ranges from about 5.8 to 2.5 or 2.0 m.y. (Menges and McFadden, 1981). (3) An alluvial "pediment gravel" unit, generally <20 m thick, that overlies unit 2 with slight angular discordance. I propose to name this unit the *Douglas Ranch Alloformation,* after a ranch in its type area, the high mesas south of Gardner Canyon. Its age probably is between 2.5 and 1 m.y. (4) The *Martinez surface.* This geomorphic surface has been described by Menges (1981) and Menges and McFadden (1981). It is a pediment-terrace surface on unit 3 that merges with the Sonoita surface along mountain margins. It is approximately the same age as or only slightly younger than the Douglas Ranch Alloformation; consequently, it seems unnecessary to name the minor erosion surface beneath unit 3. The least-eroded remnants of the Martinez surface bear very strongly developed soil profiles, typically Paleargids (Martinez soil series) with textural B horizons commonly 1.5 to 2 m thick; the clay is mostly kaolinite, which attests to the considerable age of this paleosol (younger soil profiles have little or no kaolinite) (McFadden, 1978, 1981). At altitudes below about 750 m, these paleosols are replaced by Paleorthids that lack a B horizon.

Approximate chrono-correlatives of these units appear to be as follows:

(1) An equivalent of the Sonoita surface is the "Rillito surface" in the Tucson basin-Canada del Oro Valley area (Pashley, 1966; Davidson, 1973; Menges and McFadden, 1981).

(2) Equivalents of the Gardner Canyon Alloformation appear to be the main parts of the Fort Lowell Formation in the Tucson basin (Davidson, 1973), the lower and middle members of the Saint David Formation in the upper San Pedro Valley

Figure 9. Northeastern piedmont of the Pinaleno Mountains southwest of Safford, showing the Martinez surface (M) at Frye Mesa. This geomorphic surface has a thin veneer of boulder gravel of the Douglas Ranch Alloformation, which is underlain by the much thicker and less coarse alluvial gravel and sand of the Gardner Canyon Alloformation (GC). The low mesas in the mid-distance are the moderately dissected middle Pleistocene surface (MP).

(Gray, 1967; Johnson and others, 1975), and the upper zone of the Gila Formation in the Duncan and Safford areas (Morrison, 1965a).

(3) Equivalents of the Martinez surface and the underlying Douglas Ranch Alloformation are the "Cordonnes surface" and its underlying pediment gravel in the upper Canada del Oro Valley (Menges and McFadden, 1981; my own field observations, 1972–1977), the lower part of the upper member of the Saint David Formation in the upper San Pedro Valley, and the capping of Frye Mesa, southwest of Safford (Fig. 9).

The long episode of general landscape stability while the Sonoita and Martinez surfaces developed seems to have ended chiefly because climatic oscillations became larger in amplitude with the onset of the Pleistocene and induced frequent episodes of accelerated erosion and deposition throughout this region. These climatic oscillations involved changes in both temperature and precipitation, which commonly were out of phase with each other to varying degrees. They induced climatically controlled changes in effective moisture and runoff, and are recorded by features such as erosional unconformities, variation in amount and particle size of sediment, and paleosols. Tectonism remained relatively minor in most areas. Also, through drainage was becoming established between the various intermontane basins, ending the former closed-basin regimens. Some basins became integrated much earlier than others. Some basins became integrated during the late Pliocene or earliest Pleistocene (Tucson and Sonoita-Patagonia basins) and others during the middle Pleistocene (upper San Pedro Valley, upper Gila-Safford-San Simon

Valley, Duncan-Clifton Valley, and Date Creek Valley). Two areas, central Sulfur Springs Valley and Hualapai Valley, still retain interior drainage. As a result, the uppermost part of the aggradational "basin fill" (contrasted with younger terrace-pediment mantles deposited after commencement of exterior drainage) ranges considerably in age from basin to basin and from one part to another of a single intermontane basin.

The differences in age, lithology, and stratigraphy of the exposed basin deposits result from several factors:

(1) time of development of exterior drainage from the basin;

(2) local tectonism during and subsequent to deposition of the exposed basin fill and its effect on configuration of the basin and depositional history;

(3) differences in local source materials from adjoining highlands, volcanic centers, and so forth; and

(4) distribution of alluvial, paludal, and lacustrine lithofacies at various stratigraphic levels.

An especially important study of basin-fill stratigraphy is that of the upper San Pedro Valley (Johnson and others, 1975; Figs. 10 and 11), which used magnetostratigraphic sampling and mapping, linked to intensive study of mammalian faunas at 13 sites. The basin-fill sequence studied ranges in age from within polarity epoch 5 up into the Brunhes polarity epoch, about 5 m.y. of record, including the whole Pliocene and the first 1.1 m.y. of the Quaternary. Because of its long mammalian record, this sequence has become the major frame of reference for correlating North American land mammal fauna zones with the geomagnetic

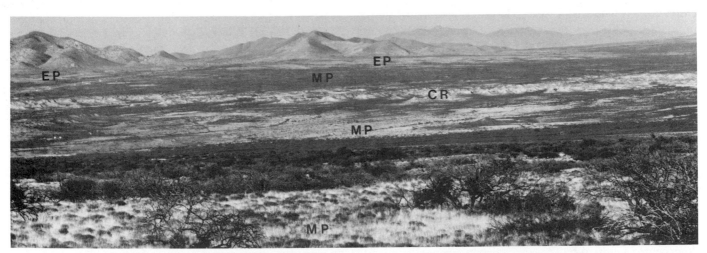

Figure 10. View eastward across the upper San Pedro Valley. Dragoon Mountains in far-middle distance, Chiricahua Mountains in far-right distance (maximum altitude of 2,985 m). EP = early Pleistocene surface, MP = middle Pleistocene surface, CR = Curtis Ranch (site of Fig. 11, in the light-toned badlands along the San Pedro River, where Miocene to middle-Pleistocene basin-fill sediments of the St. David Formation are well exposed).

Figure 11. Middle and upper members of the St. David Formation (Pliocene to middle Pleistocene). In the foreground are strata that record the Olduvai polarity event, which marks the Pliocene-Pleistocene boundary. Curtis Ranch, upper San Pedro Valley.

chronology (Lindsay and others, 1979). However, the stratigraphy of deposits mantling the various Pleistocene terraces and pediments in this area remains to be worked out in detail.

Quaternary Geomorphic History, Geomorphic Surfaces, and Alluvial and Lacustrine Deposits. Once exterior drainage became established in a basin, it ceased being a site for continuous aggradation and for lacustrine and paludal deposition; instead, dissection and removal of previously accumulated sediment were the general regime. Today, the piedmonts in the Mexican Highland section commonly are dissected 10 to >100 m, forming an intricate pattern of valleys and ridges more or less transverse to the axial stream. The areas of deep dissection provide good exposures of the "basin fill" both vertically and laterally. However, dissection by streams progressed unevenly during the Quaternary; it was punctuated by alternating episodes of downcutting followed by episodes of predominantly lateral cutting. The result is a stepped series of strath terraces that commonly grade downstream into basin-interior pediments (Morrison, 1965a, 1965b; Menges, 1981). The individual terraces and pediments typically are mantled with several meters to rarely more than 10 m of alluvial gravel, which is coarser than the underlying basin fill. Thus, the piedmonts within the intermontane basins are complex in topographic and stratigraphic detail. The older, higher piedmont surfaces are generally considerably dissected, and the younger terraces and pediments are successively less dissected.

The alternating episodes of downcutting versus relative landscape stability that produced the stepped Quaternary land surfaces in the basin interiors seem to be the result of climatic oscillations that had generally similar effects throughout this region. The fact that similar, approximately age-equivalent geomorphic/allostratigraphic units (especially the two main Pleistocene terrace/pediment sets discussed below) occur in each basin in the Mexican Highland section bespeaks a regional climatic, rather than a tectonic, control of Quaternary erosion and sedimentation. However, climate occasionally was superseded by tectonism in limited areas.

Two major sets of Pleistocene pediments and strath terraces are present, chiefly on the piedmonts of the intermontane basins; thus, they are cut mainly on basin fill; rarely do they extend onto hard rocks of the bordering mountains (Figs. 9 and 10). Each terrace-pediment set will be discussed below as an individual land surface, although locally it can be subdivided into more than one terrace or pediment, closely related in age and height above principal streams. The older terrace-pediment set, believed to be early Pleistocene, will be called here the *early Pleistocene surface.* It is preserved in widely scattered remnants along the margins of some basins (some basin margins are devoid of these remnants). This surface usually is 300 to 600 m, rarely as much as 1,000 m, above the axial stream and 30 to 150 m above principal side streams. It usually is mantled with boulder/cobble gravel, with boulders commonly a meter across. Gradients of the end-depositional surface generally are at least 1 degree and locally exceed 3 degrees. Apical portions typically are ancient alluvial fans, but distal parts

generally are pediment cappings over much less coarse basin fill. The relict soils are Paleargids or Paleorthids. The Paleargids have textural B horizons greater than 1 m thick where least eroded; generally they are considerably truncated. Above 1,050 to 1,200 m in altitude, Cca horizons are absent, but at lower altitudes considerable Cca accumulation usually is present; below about 900 m, the relict soils are mostly Paleorthids.

The younger terrace-pediment set, here called the *middle Pleistocene surface,* probably formed between the end of the Jaramillo polarity event (about 0.9 m.y. ago) and about 0.25 m.y. ago. The earlier members of the set usually are the most extensive. Thus, in terms of classic glacial units of the U.S. Midwest, the middle Pleistocene surface is mainly of Kansan and partly of Illinoian age.

The middle Pleistocene surface, seen chiefly on the piedmonts of the intermontane basins, is represented by both pediments and strath terraces, generally cut on basin fill. Pediments of this age are generally absent on resistant rocks at the mountain fronts, although narrow ones occur locally on weak bedrock. In the mountains, strath terraces coeval with this surface occur in places along the larger streams, but rarely are more than a few meters wide. Indeed, the small amount of bedrock pedimentation during the middle and late Quaternary is a significant, little-recognized feature of the erosional history of this region.

The alluvial gravel that mantles the strath terraces and pediments of the middle Pleistocene surface varies from boulder gravel near the mountain fronts to mainly cobbly pebble gravel and pebble gravel near the axial streams of the basin interiors. Typically, this alluvium is finer grained than that of the early Pleistocene surface, and it generally is coarser than the late Quaternary alluvium in any given area. This alluvial mantle bears strongly developed Haplargids (locally even Paleargids), commonly with moderately to strongly developed Cca horizons.

In the Duncan-Clifton and Upper Gila-Safford-San Simon Valleys, the alluvium on this surface locally interfingers with lacustrine deposits, mostly well-rounded pebble to cobble gravel, commonly of mixed lithology, that indicate considerable longshore transport. Shore terraces and bay bars are preserved in a few places that have been sheltered from active subsequent erosion. The distribution of shore deposits indicates that at the highest lake stage, probably during early-middle Pleistocene, the ancient lake in the Upper Gila-Safford-San Simon Valley extended from 6 km northeast of Globe to near San Simon, a distance of about 190 km, and had a maximum depth of nearly 100 m. The approximately coeval high-level lake in the Duncan-Clifton Valley extended from several km southwest of Clifton to 29 km southeast of Duncan, a distance of about 72 km, and had a maximum depth of at least 50 m. In both basins, the highest strandline varies considerably in altitude from place to place because of subsequent deformation; it ranges from 1,200 to 1,340 m in the Duncan-Clifton basin (Morrison, 1965a) and between about 1,130 and 1,300 m in the Upper Gila-Safford-San Simon Valley. Prior to these deep lakes, both basins had interior drainage. Establishment of the Gila River and exterior drainage from these basins likely came about as a result of overflow of the lake in each basin, instead of by headward erosion and capture.

In many of the intermontane basins in the Mexican Highland section, the middle Pleistocene surface and associated piedmont gravels merge downslope with coeval river terraces along the axial stream. The river terraces are mantled with pebble/cobble gravel of mixed lithology, including exotic, far-traveled rock types, that contrasts with the locally derived lithology of the alluvial gravel on the piedmonts.

Late Pleistocene and Holocene alluvium underlies the flood plains and lower terraces along the axial streams, as well as narrow strips along the larger side streams. In this region, these deposits are much less extensive than those of middle Pleistocene age, except in the few intermontane basins that have remained closed (Hualapai Valley and central Sulfur Springs Valley) or that have sluggish through drainage and, hence, aggrading central portions. Relict soils on the late Pleistocene alluvium are moderately to weakly developed Haplargids and Camborthids; on the Holocene alluvium, they are essentially all Entisols.

Eolian sand is rare. The most extensive deposits are along the eastern side of San Simon Valley near the southern end of the Whitlock Mountains.

Volcanic Fields. The Mexican Highland section has two volcanic fields in Arizona. The San Carlos field includes the early Pleistocene alkali-basalt flows of Peridot Mesa and Flatiron Mesa near San Carlos (K-Ar date from Peridot Mesa is 0.93 m.y., Shafiqullah and others, 1980, indicating a downcutting rate of approximately 72 m per m.y. at the San Carlos River), as well as the topographically higher and much larger basalt field that underlies Black Mesa and the western and central parts of Ash Flat (Bromfield and Shride, 1956; Holloway and Cross, 1978; Wohletz, 1978). The Black Mesa-Ash Flat flows have not been studied, but from their geomorphic relations they appear to be Pliocene.

The San Bernardino volcanic field occupies most of the U.S. portion of San Bernardino Valley (a southward extension of the San Simon Valley), with protrusions onto the flanks of the Chiricahua, Pedregosa, and Peloncillo Mountains (Lynch, 1978). It ends at the U.S.-Mexico border. It consists of basaltic flows and some basaltic cinders and lapilli. The oldest flows, about 3.3 m.y. old, are exposed only on the mountain flanks and predate several hundred meters of downdropping of the intermontane basin floor. Most of the exposed volcanics are between 1 and 0.27 m.y. old; however, a steam-blast eruption at Paramore Crater deposited a small amount of lapilli tuff between 100,000 and 25,000 years ago, judging from the degree of soil development on the tuff.

SURFICIAL GEOLOGY AND GEOMORPHOLOGY OF THE SONORAN DESERT SECTION

General Features

In the Sonoran Desert section of the Basin and Range province, the Basin and Range deformation ended 10.5 to 6 m.y. ago, several million years earlier than in the Mexican Highland section, and significantly high-angle faulting ceased well before the beginning of the Quaternary (Shafiqullah and others, 1980; Morrison and others, 1981a). As a consequence, landscapes in this region are much older than those in the Mexican Highland section. the mountain ranges are much more worn down and worn back from the basin-margin faults. Thus, the intermontane basins ("valleys") are wider than the bordering mountain ranges. The valleys are generally 5 to 30 km wide, whereas the mountain ranges are mostly 1 to 10 km wide. The mountain ranges tend to be shorter than those in the Mexican Highland section, mostly 5 to 30 km long, although a few are more than 50 km long. Also, the mountain ranges are generally lower; the higher peaks are between 900 and 1,675 m in altitude. Significantly, the mountain ranges are much more maturely eroded than those of the Mexican Highland section—as is demonstrated by their typically deeply embayed margins, the extensive rock pediments along their margins, and the late Miocene to early Pleistocene strath terraces cut deep into their interiors (Fig. 12). A striking, much-noted feature of this region is the dramatic change in slope between the piedmont plains and mountain fronts (pediment angle), which produces scenic mountain fronts rising boldly above the desert plains. The mountains are exceptionally rugged, with many cliffy escarpments despite their low altitude.

This whole region has exterior drainage, principally toward the Gila River but locally to the Gulf of California (in the extreme south) and to the Colorado River (along the western edge). Exterior drainage began developing from previously closed basins soon after the Basin and Range deformation ended. The Gila River was a trunk stream through the Gila Bend area and Lower Gila Valley well before 3.3 m.y. ago, as is evidenced by well-rounded alluvial gravel of mixed lithology beneath the volcanics of the Gillespie and Sentinel volcanic fields, exposed in escarpments along the present flood plain of the river (Eberly and Stanley, 1978; Shafiqullah and others, 1980). Nevertheless, the axial streams of the basins of the Sonoran Desert commonly have very low gradients, resulting in sluggish through transport of sediment and local aggradation. Consequently, the intermontane basins of this region tend to be considerably less dissected than those of the Mexican Highland section—a further cause of the strong contrast between the gently sloping piedmont plains and the rugged mountains (Fig. 13).

The Upper Miocene and Pliocene Surface. Especially prominent in the Sonoran Desert is a land surface, informally called here the upper Miocene and Pliocene surface, whose time of development partly overlaps that of the Sonoita surface of the Mexican Highland section. It started forming several million years earlier in this region (in the late Miocene), but it stopped developing about the same time as the Sonoita surface—therefore, these surfaces are not exactly time-equivalent in the two regions. The land surface in the Sonoran Desert is the product of an episode of both tectonic quiescence and climatic stability that lasted 4 to 9 m.y. in this region, prior to the climatic

Figure 12. Landsat-1 multispectral image (ERTS E 1068-17382, band 5, taken 29 Sept., 1972) entirely within the Sonoran Desert section. P = Phoenix (western part), GB = Gila Bend, A = Ajo, GR = Gila River, SR = Salt River, SV = Sentinel volcanic field, GBV = Gila Bend volcanic field, GV = Gillespie volcanic field, and AV = Arlington volcanic field.

perturbations of the Pleistocene. Equilibrium between stream downcutting and aggradation thus developed and persisted in the intermontane basins during an interval about twice as long as in the Mexican Highland section. Lateral planation predominated, and extensive pediments were formed. The prolonged development of this surface is the chief reason for the deeply embayed mountain fronts characteristic of this region, where in some cases the embayments extend entirely through a mountain range.

The upper Miocene and Pliocene surface is cut mainly on

earlier basin-fill (units I and II of Eberly and Stanley, 1978) but also on other rocks, and is preserved in places as pediments fringing the mountains and as local strath terraces in the interiors of some mountain ranges. It is especially conspicuous in the southeastern part of the Sonoran Desert, where the floors of the intermontane basins remain relatively high. This area lies between Altar Valley and the Growler and Aqua Dulce Mountains and from the Sand Tank Mountains southward past the U.S.-Mexico border. These pediments also form prominent surfaces

Figure 13. Western front of the Kofa Mountains, showing the sharp contrast between the precipitous mountain slopes and the gently sloping piedmont, and the deeply embayed valleys along the mountain front—typical of the Sonoran Desert section.

fringing various mountain ranges north of the Gila River, such as the Gila Bend, Palomas, Tank, and Kofa Mountains, and within some mountains near the Colorado River. This surface generally is dissected 5 to 15 m by Quaternary erosion. The pediments and strath terraces are mantled in places by approximately coeval cobble to boulder gravel, but in many places they are stripped bedrock. Where least eroded they bear extremely well developed relict paleosols, most commonly Paleorthids (with petrocalcic horizons 1.5 to >2 m thick), but locally Paleargids bordering and within the higher mountains.

Quaternary Geomorphic Surfaces and Alluvial Deposits

The two main sets of Pleistocene pediment/strath terrace surfaces, lower Pleistocene and middle Pleistocene, that occur in the Mexican Highland section are also the main sets in the intermontane basins of the Sonoran Desert. The lower Pleistocene surface is preserved only in remnants close to the mountains, in places extending into the mountains as strath terraces. It typically is 5 to 10 m below the late Miocene-Pliocene surface, where the latter is present. It is commonly mantled with coeval alluvial

gravel, and it bears generally Paleorthids, locally Paleargids, as relict paleosols.

The middle Pleistocene surface is by far the most extensive. In many intermontane basins, it covers nearly all the piedmont plains between the mountains and the lowlands along the axial stream. It typically is 5 to 10 m below the lower Pleistocene surface (where the latter is present). It is mantled with approximately coeval alluvial gravel, 3 to 10 m thick, that bears strongly developed Haplargids and Calciorthids (locally Paleargids and Paleorthids).

In addition to these surfaces on the piedmonts, Pleistocene terraces occur along the major streams. The only modern studies are those of the terraces of several rivers where they cross from the Mexican Highland section into the Phoenix and Casa Grande basins (Péwé, 1978). (This is a rather unique area where middle and late Quaternary tectonism has intermittently superseded climate as the fundamental control of erosion and sedimentation.) Three Pleistocene terraces are present along the Salt River. They converge downstream and disappear beneath Holocene alluvium. Between the mountain front (Granite Reef Dam) and the Scottsdale-Tempe area, the highest terrace, of early Pleistocene or latest Pliocene age, grades from 72 to 15 m above the river; the middle, most prominent terrace, probably early-middle Pleistocene, grades from 46 to 3 m; the lowest terrace, likely late Pleistocene, is 15 m above the river at the mountain front and disappears beneath Holocene alluvium before reaching the Scottsdale-Tempe area. Similar sets of downstream-converging terraces occur along the Lower Verde, Agua Fria, and Gila Rivers (each with 3 Pleistocene terraces) and along Queen Creek (2 terraces). Those of the Agua Fria and Gila Rivers and Queen Creek disappear beneath Holocene alluvium within 40 km of the mountain front. Thus, Pleistocene terraces are buried in the lower parts of the Phoenix and Casa Grande basins, downstream to just below the junction of the Gila and Hassayampa Rivers. Such downstream convergence of terraces would seem to indicate that tectonic perturbation, if present, would be either (1) upstream from the reach of convergence, as Péwé (1978) suggests, or (2) downstream from this reach, perhaps due to partial damming of the Gila River at the Arlington-Gila Bend reach. Farther down the Gila, from Arlington nearly to Yuma, presumably undeformed remnants of Pleistocene river terraces are common. Here, the height of the highest middle Pleistocene terrace above the river ranges from 36 m several km west of Gila Bend to 20-25 m in the western part of Lower Gila Valley.

Terraces of the Santa Cruz River are somewhat different. Before the Santa Cruz enters the Sonoran Desert, within 32 km above and below Tucson, scattered remnants of an important lower-middle Pleistocene terrace, locally displaced by faults, are 12 to 46 m above the river. Just above Marana (approximately the boundary between the Mexican Highland and Sonoran Desert sections), this terrace is cut off by a fault, but coeval river gravel continues downstream in the subsurface (it was an important aquifer, now largely depleted; M. E. Cooley, personal commun., 1973).

Eolian Sand Deposits

Eolian sand is common in the lower-altitude basins, especially in the southwestern part of the Sonoran Desert. It underlies most of the Yuma Desert and large areas in the Lechuguilla Desert, Tule Desert, Mohawk Valley, Growler Valley, San Cristobal Valley, The Great Plain, Vamori Plain, and Cactus Plain (southeast of Parker). In these areas, it occurs mainly as transverse and seif dunes, locally as barchans, and grades to coppice dunes, sand sheets, and cover sands (mixed eolian, colluvial, and alluvial sand). Essentially all the exposed eolian deposits appear to be of late Pleistocene and Holocene age. Tiny, scattered eolian deposits, such as coppice dunes and small transverse dunes, occur in many other places in the basin lowlands, such as the Phoenix and Casa Grande areas.

The marked increase in abundance of eolian sand to the southwest suggests that principal sources were the flood plain and delta of the Colorado River, as well as the Desierto Grande and Desierto del Altar, in Sonora. Another source, both for these sand deserts and ultimately for the southerly occurrences in Arizona,

may have been the bed of the northern end of the Gulf of California, which was intermittently exposed when sea level was lowered glacioeustatically during Pleistocene glaciations. Curray and Moore (1964) report that about 17,600 years ago, during the last pleniglacial, sea level in the Gulf of California was lowered 124 m.

Volcanic Fields

Four of the five volcanic fields in the Sonoran Desert are of late Pliocene age: the Sentinel, Gila Bend, Gillespie, and Arlington volcanic fields. These are separate but so close together that they are sometimes classed as a single field. They are similar and consist entirely of basalt flows (probably single flows from a given vent) and low volcanic cones, all about the same age. Their surfaces are much weathered, bear very strongly developed Paleorthids, and are extensively mantled with loessial-colluvial silt and fine sand. More than 20 K-Ar age determinations have been made, but some determinations are questionable. Averages of the more reliable ones indicate that the main flows in each field are about 3.2 or 3.3 m.y. old. However, somewhat less caliche development, weathering, and loessial accumulation on the Gillespie field suggest that it may be somewhat younger than the other fields.

A tiny portion of the huge Pinacate volcanic field extends across the U.S.-Mexico border into Arizona. This portion consists of an upper Pleistocene basalt flow and upper Pleistocene or Holocene(?) cinders and lapilli.

SURFICIAL GEOLOGY AND GEOMORPHOLOGY OF THE LOWER COLORADO RIVER

A narrow zone along the Colorado River is a special case, quite different from the rest of the Mexican Highland and Sonoran Desert sections. Downstream from Lake Mead, the Colorado River flows perpendicular to the structural grain, crossing several ranges and intervening basins and entering the Sonoran Desert through a gorge below the mouth of the Bill Williams River. The lower Colorado River came into existence very early in the Pliocene, after deposition of the Bouse Formation (in salt to brackish water of the ancestral Gulf of California) had ended shortly after 5.47±0.20 m.y. ago, and was well established by 3.8 m.y. ago (Damon and others, 1978). The river probably was superposed on the upper Miocene and Pliocene erosion surface, for there is no evidence that this river was ponded and thus became established by overflow of successive intermontane basins (Metzger and others, 1973).

The piedmont surfaces and deposits of this zone partly resemble those of the Sonoran Desert, but also differ sharply. Their degree of dissection resembles that common in the Mexican Highland section. On the upper piedmonts, the various primary erosional-depositional surfaces typical of the main Sonoran Desert can be seen, including the upper Miocene-Pliocene and the lower Pleistocene surfaces exposed as remnants in the basin ranges and along their borders, and an extensive middle Pleistocene surface. However, the late Pliocene and Quaternary depositional history of the lower Colorado River was much affected by local tectonism, including isostatic perturbations of the delta of this river, and by glacioeustatic changes in sea level, caused by Pleistocene glaciations. This history is so complex that its details are still uncertain. The widest range in altitude of deposition of Colorado River gravels is in the Yuma area: from 177 m above sea level on Upper Mesa, south of Yuma, and 226 m above sea level on the southwest flank of the Chocolate Mountains to more than 1,000 m below sea level at the Mexican border, indicating subsidence during deposition (Olmsted and others, 1973). Upstream, in the Cibola-Blythe-Parker basin, Colorado River deposits extend from about 100 m below sea level (near Blythe) to as much as 200 m above sea level south of Parker (Metzger and others, 1973). The older alluvial deposits, of middle to late Pliocene age, are faulted and warped in places, but the Quaternary deposits are mostly undeformed. Especially prominent among the younger deposits are two sets of terraces. The higher and older set is 35±8 m above the Colorado River in the Cibola-Blythe-Parker basin and bears a strongly developed soil profile (its Cca horizon typically is Stage 3 to 4), indicating a middle Pleistocene age. The younger prominent terrace set is 5 to 10 m lower, bears a soil profile with Stage 2 to 2½ Cca development, and hence probably is no older than the last interglacial. Narrow, somewhat lower terraces also occur locally that bear more weakly developed soil profiles, suggestive of Wisconsinan to early Holocene age.

CONCLUSIONS ON QUATERNARY EROSION AND SEDIMENTATION IN ARIZONA

Arizona's volcanic fields are much better understood, in terms of their Pliocene and Quaternary stratigraphy and chronology, than any of the state's alluvial areas. This is largely due to a two-decade program of K-Ar dating of Arizona's Cenozoic volcanics at the Laboratory of Isotope Geochemistry, Department of Geosciences, University of Arizona, under the direction of Paul E. Damon. In many cases, dated or well-correlated volcanic units provide chronologic benchmarks for determining the late Cenozoic history. Nevertheless, many details of the sedimentary history still are poorly understood and poorly correlated through various parts of the state.

My general conclusions on the history of Quaternary erosion and sedimentation in Arizona are as follows.

On the Colorado Plateau, details of Quaternary erosional-depositional history are limited and ambiguous because the alluvial record is very meager (due to more or less continuous regional downcutting) and has little chronologic control.

In the Mexican Highland and Sonoran Desert sections of the Basin and Range province, two primary erosion-deposition cycles (that produced the lower Pleistocene and middle Pleistocene pediment-terrace surfaces) are evident. These cycles are manifested throughout the Basin and Range province, with its many separate intermontane basins that have big differences in altitude,

relief, climate, and local drainage history. Therefore, these cycles surely were controlled by climatic change, not by tectonism. Yet, their frequency is about a tenth of the total number of glacial cycles (20 or so) during the Pleistocene, as demonstrated by the oxygen-isotope data from deep-sea cores. Evidently the climatic changes that produced the primary erosion-deposition cycles in Arizona were megacycles, considerably longer than individual glacial cycles. The middle Pleistocene megacycle that produced the dominant pediment-terrace set in the intermontane basins lasted at least several hundred thousand years, and probably half a million or more years. It may correlate with all or much of the oxygen-isotope sequence from stages 20 to 8.

Why have the primary erosion-deposition cycles affecting the Arizona landscape danced to so slow a rhythm? Is it because Arizona lies in the subtropical arid zone, and, through the course of the Pleistocene, climatic fluctuations were buffered relative to those in more northerly zones? Did the buffering result in generally less amplitude of changes in precipitation and temperature and tend to smooth out the climatic perturbations so that only the grosser ones (the megacycles) left significant records on the landscape—in effect, causing a series of geomorphic mega-thresholds to operate in forming the Arizona landscapes?

Two additional observations should be made: (1) Correlations of episodes of vertical or lateral fluvial erosion or of landscape stability and soil-profile development with particular parts of glacial-interglacial or other climatic cycles should be made cautiously. Specifically, episodes of landscape stability and maximum soil-profile development do not necessarily correlate with glacials, nor do times of maximum erosion and aggradation necessarily correlate with interglacials, as some workers have suggested (because they interpret the glacials as having been cool-wet and the interglacials as warm-dry—probably an over simplification). (2) The present is not a true key to the past, in terms of the rates of activity of various geomorphic processes through Pliocene and Quaternary time. For example, the last important pedimentation of resistant rocks took place in the Pliocene. Conditions for significant hard-rock pedimentation do not exist now in Arizona, nor have they existed during most of the Quaternary.

RECOMMENDATIONS FOR RESEARCH ON PLIOCENE AND QUATERNARY STRATIGRAPHY AND GEOMORPHOLOGY IN ARIZONA

The best-exposed, fullest sequences of Pliocene to lower and middle Pleistocene strata in Arizona are in parts of the following intermontane basins: (1) middle and upper San Pedro Valley, (2) Safford (Upper Gila)-lower San Simon Valley, (3) Duncan-Clifton Valley, (4) Date Creek-Bullard Wash Valley, and (5) several areas along the lower Colorado River from Lake Mohave to Cibola. These areas have large, almost untapped, reservoirs of stratigraphic and climatic data. Few other areas in western North America have equivalent records for this time span that do not require subsurface exploration. Study of these areas should be multifaceted to yield the greatest and least controversial amount of information, and should include:

(1) detailed, large-scale mapping of lithostratigraphic and allostratigraphic units, including key beds, unconformities, key paleosols, and fossil localities;

(2) measurement of stratigraphic sections and cross sections in detail at representative sites;

(3) magnetostratigraphic studies, both reconnaissance and in detail at key sites;

(4) analysis and dating of volcanic ash beds;

(5) mapping the various geomorphic surfaces and their veneering deposits and relating the latter to the basin-fill stratigraphy;

(6) pedologic studies, including analysis of particle size, clay mineralogy, and key chemical and physical constituents of relict soil profiles on the various geomorphic surfaces of buried paleosols; and

(7) dating key beds, soil profiles, and so forth, by whatever means applicable (for example, radiocarbon, K-Ar, fission track, thermoluminescence, magnetostratigraphy, and uranium-series).

These recommendations emphasize the need for intensified, stratigraphically oriented studies. I believe that, in landscapes that preserve so much of the far past as those of Arizona, their history will be revealed only to those who follow such a path.

NEOTECTONIC (LATE PLIOCENE AND QUATERNARY) FAULTS

Recently a comprehensive map of late Pliocene and Quaternary faults (<3.5 m.y. old) in the state of Arizona was completed at a scale of 1:500,000 (Morrison and others, 1981a, 1981b). This map is based chiefly on interpretation of U-2 air photos, but is supported by our own extensive ground and aerial reconnaissance.

Means of Identification of Neotectonic Faulting

We used "hard" evidence of proven offset of upper Pliocene and Quaternary strata wherever it could be obtained, from published maps or our own field observations, but such evidence is comparatively rare. Thus, we were obliged to use various indirect means to determine most of the suspected faults. We relied heavily on geomorphic analysis to identify suspected surface offsets. Our analysis had to be multifaceted because we attempted to distinguish faults that range considerably older than those usually classed as "neotectonic." The chief kinds of indicators that we used are:

(1) *Scarp analysis.* Studies that attempt to identify young faults by means of ground-breaking scarps seen on air photos or on the ground generally concentrate on scarps so young as to be clearly evident because of little erosional modification. Such scarps are mostly less than 100,000 years old and never more than a few hundred thousand years old, unless in resistant rocks such as basalt. We looked also for older fault scarps, which, being

more eroded (with lower scarp angle and a greater retreat from the fault trace), are less obvious and more ambiguous than the younger scarps.

Most of the fault scarps we identified are on piedmonts within intermontane basins or within or adjoining upper Cenozoic volcanic fields. They commonly separate discordant land surfaces and are associated with drainage anomalies (see below). Those in the intermontane basins are fairly obvious and unambiguous where they are oriented across the local drainage. Where they parallel the drainage, they are more ambiguous, and other criteria must be applied to distinguish stream-eroded from tectonic scarps.

(2) *Discordant land surfaces* signify an aligned discontinuity between geomorphic depositional or erosional surfaces of Pliocene or Quaternary age (mostly on piedmonts, pediments, or volcanic fields) that more likely is due to faulting than to normal erosion or deposition. The land-surface can either be the same age or different ages; each surface is defined by accordant ridge summits, and generally they are separated by a scarp and (or) a drainage anomaly. A discordance that crosses the local drainage is highly indicative of faulting if it is at least several kilometers long and continues across several different drainages. A marked discordance that parallels the local drainage usually is less ambiguous than the stream-eroded scarp along the discordance.

Some land-surface discordances occur along the axial streams of intermontane basins. One piedmont surface is higher (that is, graded to a higher elevation than the opposite piedmont with respect to the axial stream) and generally is the more deeply dissected.

(3) *Drainage anomalies* are various kinds of abnormal stream alignments manifested by (a) abrupt deflection of stream direction (very suggestive if repeated by more than one stream along the same alignment); (b) abrupt aligned change in the drainage density; (c) discontinuous gullies or minor valleys in the same alignment across more than one large stream; (d) aligned ponding or semiponding, commonly against the uplifted member of a pair of discordant land-surfaces; and (e) long, continuous straight or gently curving alignment of the same stream (suggestive only if this is the boundary between discordant land surfaces).

(4) *A lithologic anomaly* in Quaternary or Pliocene rocks that cannot be explained by normal changes in lithofacies or by nontectonic erosion or deposition, as shown by abrupt change in particle size, in consolidation, induration, stream density, stream pattern, or in ridge height or width. This indicator is useful where distinctive marker beds or units are absent and stratigraphic offset cannot be determined, but it is generally more applicable to basin-fill sediments (Pliocene and older) than to Quaternary piedmont deposits.

The many kinds of evidence we were obliged to consider in order to identify faults that are much older than those usually considered to be "young" or "neotectonic" required classifying the suspected faults as to various degrees of probability: "proven," "probable," and two categories of "very possible," either very

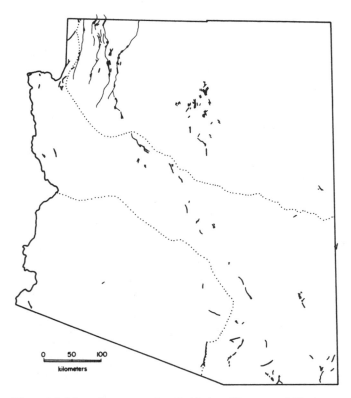

Figure 14. Map of proven and probable late Pliocene and Quaternary faults (heavy lines) in Arizona. Only a few "very possible" faults are included that are directly associated with proven and probable faults. The dotted lines show physiographic boundaries (see Fig. 1).

possible displacement of deposits or landforms <3.5 m.y. old, or very possible displacement younger than 3.5 m.y. along a proven or probable fault in pre-upper Pliocene rocks (Fig. 14).

Evidence for Age of Faulting

We used the following kinds of evidence to determine the age of neotectonic fault displacement:

(1) *Offset of strata of known Pliocene or Quaternary age.* The strata are either reliably dated by geochemical or paleomagnetic methods (rarely applicable) or are reliably correlated with dated units (used in volcanic fields with stratigraphic units dated by K-Ar).

(2) *Stratigraphic analysis* of surficial deposits, including soil profiles, on both sides of the fault (used in connection with field studies in some cases).

(3) *Scarp evidence:* (a) freshness of the fault scarp, specifically, the proportion of width of the debris slope to the sum of the wash slope and the crestal zone; (b) scarp angle, compared with the scarp height, and total profile; (c) straightness of the scarp; (d) strath-terrace analysis on streams crossing the scarp; and (e) analysis of alluvial fan-piedmont surfaces and stratigraphy on both sides of the scarp.

(4) *Geomorphic land-surface evidence* from dated or reliably correlated piedmont surfaces, volcanic surfaces, erosion sur-

faces, or pediments that are displaced by neotectonic faults. By far the commonest means of dating displacements is our own correlation of various Pliocene-Quaternary deposits, land surfaces, and allostratigraphic units from our mapping of these features throughout the state. An important aid for this correlation is the relative degree of development of the most ancient and uneroded soil profiles on the various land surfaces.

RESULTS AND CONCLUSIONS ON NEOTECTONIC FAULTS

About 400 neotectonic faults or fault segments have been mapped in Arizona whose probability ratings range from "proven" to "suspected, very possible." About one third of these are in the "proven" and "probable" categories.

All of these faults are normal, high-angle, and (within our means of determination) entirely or largely dip-slip. They have the same orientations as those of the Basin and Range deformation *sensu strictu,* mostly northwest, west-northwest, north-northwest, north, and northeast. They seem to be reactivated earlier faults. Generally they are marked by scarps 3 to 20 m high (some scarps are >40 m high), with maximum scarp-slope angle of 3 to 25 degrees (most commonly 6 to 15 degrees) (Morrison and others, 1981a, 1981b; Menges and others, 1982).

Twelve areas, scattered in a broad belt northwest to southeast across Arizona, plus the Yuma area, have the most neotectonic faults (Fig. 15). Neotectonic faults are essentially lacking or not discernible in: (1) the Colorado Plateau east and north of the San Francisco and White Mountain volcanic fields; (2) the transitional zone of the Mexican Highland section; and (3) the Sonoran Desert east of the Yuma basin.

In the Basin and Range province, most of the neotectonic faults are within the intermontane basins. Few faults seem to be in the mountain blocks, although high relief and scarcity of Quaternary deposits adversely affect the recognition of such faults. Basin-margin faults (produced or reactivated by the Basin and Range deformation) show neotectonic displacement only locally, and many lack such displacement entirely, as shown by lack of offset of lower Pleistocene to upper Pliocene deposits and land surfaces.

Recent detailed field studies by C. M. Menges, P. A. Pearthree, and L. Mayer (Pearthree and others, 1983) require some modification of earlier estimates (Morrison and others, 1981a, 1981b) of the latest surface rupture, recurrence intervals, and the magnitudes of earthquakes that are recorded by surface displacements in Arizona. These field studies concentrated on the neotectonic faults previously classed as "proven" or "probable." They involved measuring fault-scarp profiles at several sites on each fault, computerized statistical analysis of each set of profiles, and observations on soil stratigraphy and alluvial/colluvial stratigraphy on both sides of the fault. However, Mayer's (1982) warning about the pitfalls and reliability of age-estimates from fault-scarp analysis is pertinent. Only after trenching across a suspected fault rupture and after competent stratigraphic (includ-

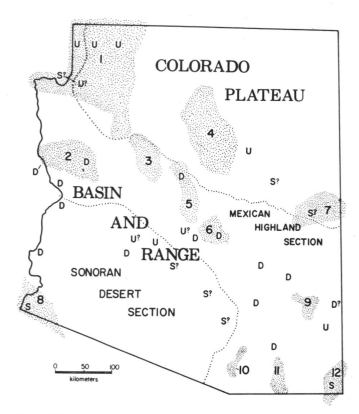

Figure 15. State of Arizona showing areas with concentrations of neotectonic faults (stippled pattern) and areas of broad late Pliocene-Quaternary uplift (U), subsidence (S), or deformation (D) (see Morrison and others, 1981a, for details). Dotted lines indicate boundaries of physiographic divisions.

ing soil-stratigraphic) analysis of the exposure can reliable assessments be made of the age of latest displacement and (possibly) of recurrence interval.

The last surface ruptures on most neotectonic faults in Arizona were in the early to early-late Pleistocene, for example, more than 100,000 years ago. However, a few faults have surface displacements as young as late Pleistocene and perhaps early Holocene. These young faults are in southeastern Arizona near Portal and southwest of Safford, also near Yuma, and in Chino Valley; also, more than 50 of them are in northwestern Arizona near the edge of the Colorado Plateau.

According to Pearthree and others (1983), recurrence intervals between surface ruptures by individual faults are 10^5 years in central Arizona and in the Lake Mead area, but they are 10^4 years in the Yuma area and in northwestern Arizona near the edge of the Colorado Plateau. Pearthree and others (1983) estimate regional recurrence intervals during the past 15,000 to 20,000 years to be 700 to 2,000 years in northwest Arizona, 2,500 to 3,000 years in southwestern Arizona, and at least 15,000 years in the Lake Mead area. They also conclude that the rate of surface rupture in central and southeastern Arizona has been at least four times greater during the past 15,000 years than it was during the previous 50,000 to 100,000 years.

Estimated energies released by the larger single surface-displacement events (based mainly on scarp height and length) indicate earthquake magnitudes of 6.5 and greater (Sbar and others, 1980). The 1887 earthquake in northern Sonora, with its epicenter about 40 km south of the U.S.-Mexico border, had an estimated magnitude of 7.4, determined from seismic moment (Sbar and others, 1981). According to Pearthree and others (1983), the magnitudes of the earthquakes that produced the surface ruptures (calculated from geologic moment estimates) range from 6.3 to 7.4 in northwestern Arizona to 5.8 to 6.5 in central Arizona, 6.1 to 7.3 in southeastern Arizona, and 6.0 to 6.6 in southwestern Arizona.

From these data, it would seem that the probability of a severe earthquake occurring in any given year in Arizona is low. However, it is important to remember that we identified only surface displacements. They do not provide a complete history of past earthquakes because many earthquakes do not cause surface rupture. Moreover, earthquakes that originate in areas bordering Arizona may affect parts of Arizona, albeit progressively attenuated with distance from the epicenter. (For example, Yuma would be more affected than Phoenix by an earthquake whose epicenter was in the Imperial Valley.) Consequently, from our maps alone it is not possible to make accurate estimates of seismic risk.

ACKNOWLEDGMENTS

I am indebted greatly to two people who encouraged me to develop many of the ideas expressed in this paper: Maurice Cooley (while we jointly worked on a NASA-financed project in 1973) and Chris Menges (who was my associate from 1979 to 1982 while we prepared the neotectonic and Quaternary geologic maps of Arizona, funded by U.S.G.S.). I also profited greatly from contacts over many years with Paul Damon, William Bull, Wesley Peirce, Vance Haynes, Robert Scarborough, Dan Lynch, and "Shafi" Muhammad Shafiqullah. Bull, Cooley, Damon, Menges, and Peirce made very helpful critiques of early versions of this paper. But "the buck stops here"—I accept responsibility for the statements made herein.

REFERENCES CITED

Anderson, R. E., Longwell, C. R., Armstrong, R. L., and Marvin, R. F., 1972, Significance of K-Ar ages of Tertiary rocks from the Lake Mead region, Nevada-Arizona: Geological Society of America Bulletin, v. 83, p. 273–288.

Bierce, A., 1911, The Devil's dictionary: New York, Albert and Charles Boni, Inc. (reprinted 1948 by Forum Books, World Publishing Co., Cleveland), 376 p.

Bromfield, C. S., and Shride, A. F., 1956, Mineral resources of the San Carlos Indian Reservation: U.S. Geological Survey Bulletin 1027-N, p. 618–691, scale 1:250,000.

Budden, R. T., 1975, The Tortolita-Santa Catalina Mountain complex [M.S. thesis]: Tucson, University of Arizona, 133 p.

Bull, W. B., 1979, Threshold of critical power in streams: Geological Society of America Bulletin, Part I, v. 90, p. 453–464.

Childs, O. E., 1948, Geomorphology of the valley of the Little Colorado River, Arizona: Geological Society of America Bulletin, v. 59, p. 353–388.

Cooley, M. E., and Akers, J. P., 1961a, Ancient erosion cycles of the Little Colorado River, Arizona and New Mexico, *in* Short papers in the geologic and hydrologic sciences: U.S. Geological Survey Professional Paper 424-C, p. 244–248.

——1961b, Late Cenozoic geohydrology in the central and southern parts of Navajo and Apache Counties, Arizona: Arizona Geological Society Digest, v. 4, p. 69–77.

Cooley, M. E., Harshbarger, J. W., Akers, J. P., and Hardt, W. F., 1969, Regional hydrogeology of the Navajo and Hopi Indian Reservations, Arizona, New Mexico, and Utah: U.S. Geological Survey Professional Paper 521-A, 58 p.

Cooley, M. E., and Wilson, A., 1968, Canyon cutting in the Colorado River system, *in* Fairbridge, R. W., ed., Encyclopedia of geomorphology: New York, Reinhold, p. 98–102.

Curray, J. R., and Moore, D. G., 1964, Pleistocene deltaic progradation of continental terrace, Costa de Nayarit, Mexico, *in* Marine geology of the Gulf of California—a symposium: American Association of Petroleum Geologists Memoir 3, p. 208.

Damon, P. E., 1971, The relationship between late Cenozoic volcanism and tectonism and orogenic-epeirogenic periodicity, *in* Turekian, K. K., ed., The late Cenozoic glacial ages: New Haven and London, Yale University Press, p. 15–35.

Damon, P. E., and Bikerman, M., 1964, Potassium-argon dating of post-Laramide plutonic and volcanic rocks within the Basin and Range province of southeastern Arizona and adjacent areas: Arizona Geological Digest, v. 7, p. 63–78.

Damon, P. E., Laughlin, A. W., and Percious, J. K., 1967, Problems of excess argon-40 in volcanic rocks, *in* Radioactive dating and methods of low-level counting: Vienna, International Atomic Energy Agency, p. 463–481.

Damon, P. E., Shafiqullah, M., and Leventhal, J. S., 1974, K-Ar chronology for the San Francisco volcanic field and rate of erosion of the Little Colorado River, *in* Karlstrom, T.N.V., and others, eds., Geology of northern Arizona with notes on archaeology and paleoclimate, Part 1, Regional studies: Geological Society of America Guidebook, Rocky Mountain Section Meeting, Flagstaff, p. 221–235.

Damon, P. E., Shafiqullah, M., and Scarborough, R. B., 1978, Revised chronology for critical stages in the evolution of the lower Colorado River [abs.]: Geological Society of America Abstracts with Programs, v. 10, p. 101–102.

Davidson, E. S., 1973, Geohydrology and water resources of the Tucson basin, Arizona: U.S. Geological Survey Water-Supply Paper 1939-E, 80 p.

Davis, W. M., 1901, An excursion to the Grand Canyon of the Colorado: Harvard Museum of Comparative Zoology Bulletin 38, p. 106–201.

Dutton, C. E., 1882, Tertiary history of the Grand Canyon district: U.S. Geological Survey Monograph 2, 264 p.

Eberly, L. D., and Stanley, T. B., 1978, Cenozoic stratigraphy and geologic history of southwestern Arizona: Geological Society of America Bulletin, v. 89, p. 921–940.

Gray, R. S., 1967, Petrology of the upper Cenozoic non-marine sediments in the San Pedro Valley, Arizona: Journal of Sedimentary Petrology, v. 37, p. 774–789.

Gregory, H. E., 1917, Geology of the Navajo country: U.S. Geological Survey Professional Paper 93, 161 p.

Hamblin, W. K., 1974, Late Cenozoic volcanism in the western Grand Canyon, *in* Breed, W. J., and Best, M. G., eds., Geology of the Grand Canyon: Flagstaff, Museum of Northern Arizona and Grand Canyon Natural History Association, p. 142–169.

Holloway, J. R., and Cross, C., 1978, The San Carlos alkaline rock association, *in* Guide to the geology of central Arizona: Arizona Bureau of Geology and Mineral Technology Special Publication 2, p. 171–173.

Johnson, N. M., Opdyke, N. D., and Lindsay, E. H., 1975, Magnetic polarity stratigraphy of Pliocene-Pleistocene terrestrial deposits and vertebrate faunas, San Pedro Valley, Arizona: Geological Society of America Bulletin, v. 86, p. 5–12.

Laughlin, A. W., Brookins, D. G., Damon, P. E., and Shafiqullah, M., 1979, Late Cenozoic volcanism of the central Jemez zone, Arizona-New Mexico: Iso-

chron/West, no. 25, p. 5–8.

Laughlin, A. W., Damon, P. E., and Shafiqullah, M., 1980, New K-Ar dates from the Springerville volcanic field, central Jemez zone, Apache County, Arizona: Isochron/West, no. 29, p. 3–4.

Lindsay, E. H., Johnson, N. M., and Opdyke, N. D., 1979, Preliminary correlation of North American land mammal ages and geomagnetic chronology, *in* Studies on Cenozoic paleontology and stratigraphy in honor of Claude W. Hibbard: University of Michigan Papers on Paleontology 12, p. 111–119.

Luedke, R. G., and Smith, R. L., 1978, Map showing distribution, composition, and age of late Cenozoic volcanic centers in Arizona and New Mexico: U.S. Geological Survey Miscellaneous Investigations Map I-1091-A, scale 1:1,000,000.

Lynch, D. J., 1978, The San Bernardino volcanic field of southeastern Arizona, *in* Callender, J. F., and others, eds., Land of Cochise: New Mexico Geological Society, 29th Field Conference, Guidebook, p. 261–268.

Mayer, L., 1982, Constraints on morphologic-age estimation of Quaternary fault scarps based on statistical analysis of scarps in the Basin and Range province, Arizona, and northeastern Sonora, Mexico [abs.]: Geological Society of America Abstracts with Programs, v. 14, p. 213.

McCann, F. T., 1938, Ancient erosion surfaces in the Gallup-Zuni area, New Mexico: American Journal of Science, 5th series, v. 36, p. 260–278.

McFadden, L. D., 1978, Soils of the Canada del Oro Valley, southern Arizona [M.S. thesis]: Tucson, University of Arizona, 116 p.

——1981, Quaternary evolution of the Canada del Oro Valley, southeastern Arizona: Arizona Geological Society Digest, v. 13, p. 13–19.

McKee, E. D., Hamblin, W. K., and Damon, P. E., 1968, K-Ar age of lava dam in Grand Canyon: Geological Society of America Bulletin, v. 79, p. 133–136.

Menges, C. M., 1981, The Sonoita Creek Basin: Implications for late Cenozoic tectonic evolution of basins and ranges in southeastern Arizona [M.S. Thesis]: Tucson, University of Arizona, 239 p.

Menges, C. M., and McFadden, L. D., 1981, Evidence for a latest Miocene to Pliocene transition from Basin-Range tectonic to post-tectonic landscape evolution in southeastern Arizona: Arizona Geological Society Digest, v. 13, p. 151–160.

Menges, C. M., Pearthree, P. A., and Calvo, S., 1982, Quaternary faulting in southeast Arizona and adjacent Sonora, Mexico [abs.]: Geological Society of America Abstracts with Programs, v. 14, p. 215.

Merrill, R. K., and Péwé, T. L., 1977, Late Cenozoic geology of the White Mountains, Arizona: Arizona Bureau of Geology and Mineral Technology Special Paper 1, 65 p.

Metzger, D. G., Loeltz, O. J., and Irelan, B., 1973, Geohydrology of the Parker-Blythe-Cibola area, Arizona and California: U.S. Geological Survey Professional Paper 486-G, 130 p.

Morrison, R. B., 1965a, Geologic map of the Duncan and Canador Peak quadrangles, Arizona and New Mexico: U.S. Geological Survey Miscellaneous Investigations Map I-442, scale, 1:48,000.

——1965b, Quaternary surfaces and associated deposits in Duncan Valley, Arizona-New Mexico (Figure 9-26), *in* International Association for Quaternary Research (INQUA), VII Congress, 1965, Guidebook, Field Conference H (Southwestern Arid Lands), p. 68.

——1982, Maps of Quaternary geology of Arizona (Gila River, Los Angeles, Mt. Whitney, Grand Canyon, and Sonora 1:1,000,000-scale quadrangles— Arizona portions only): On file at U.S. Geological Survey, Federal Center, Denver, Colorado, and at Arizona Bureau of Geology and Mineral Technology, Tucson, Arizona.

Morrison, R. B., Menges, C. M., and Lepley, L. K., 1981a, Neotectonic maps of Arizona, in Stone, C., ed.: Arizona Geological Society Digest, v. 13, p. 179–183.

——1981b, Map of late Pliocene to Quaternary faults in Arizona (scale 1:500,000): On file at U.S. Geological Survey, Menlo Park, California, and at Arizona Bureau of Geology and Mineral Technology, Tucson, Arizona.

North American Commission on Stratigraphic Nomenclature, 1983, North American stratigraphic code: American Association of Petroleum Geologists Bulletin, v. 67, p. 841–875.

Olmsted, F. H., Loeltz, O. J., and Irelan, B., 1973, Geohydrology of the Yuma area, Arizona and California: U.S. Geological Survey Professional Paper 486-H, 227 p.

Pashley, E. F., 1966, Structure and stratigraphy of the central, northeastern, and eastern parts of the Tucson Basin, Pima County, Arizona: [Ph.D. thesis]: Tucson, University of Arizona, 86 p.

Pearthree, P. A., Menges, C. M., and Mayer, L., 1983, Distribution, recurrence intervals, and estimates of magnitude of Quaternary faulting in Arizona [abs.]: Geological Society of America Abstracts with Programs, v. 15, p. 417.

Peirce, H. W., Damon, P. E., and Shafiqullah, M., 1979, An Oligocene(?) Colorado Plateau edge in Arizona, *in* McGetchin, T. R., and Merrill, R. B., eds., Plateau uplift, mode and mechanism: Tectonophysics, v. 61, p. 1–24.

Péwé, T. L., 1978, Terraces of the Lower Salt River Valley in relation to the late Cenozoic history of the Phoenix basin, Arizona (and field trip log), *in* Burt, D. M., and Péwé, T. L., eds., Guidebook to the geology of central Arizona: Arizona Bureau of Geology and Mineral Technology Special Paper 2, p. 1–45.

Péwé, T. L., and Updike, R. G., 1976, San Francisco Peaks, a guidebook to the geology (second edition): Flagstaff, Museum of Northern Arizona, 80 p.

Sbar, M. L., DuBois, S. M., and Bull, W. B., 1980, Preliminary assessment of seismic hazards in Arizona [abs.]: Earthquake Notes, v. 50, p. 6.

Sbar, M. L., DuBois, S. M., and Natali, S. G., 1981, Analysis of the 1887 Northern Sonora earthquake and microearthquake study of the epicentral region [abs.]: Geological Society of America Abstracts with Programs, v. 13, p. 104–105.

Scarborough, R. B., and Peirce, H. W., 1978, Late Cenozoic basins of Arizona, *in* Callender, J. F., Wilt, J. C., and Clemons, R. E., eds., Land of Cochise: New Mexico Geological Society, 29th Field Conference, Guidebook, p. 253–259.

Shafiqullah, M., Damon, P. E., Lynch, D. J., and Kuck, P. H., 1978, Mid-Tertiary magnetism in southeastern Arizona, *in* Callender, J. F., and others, eds., Land of Cochise: New Mexico Geological Society Guidebook, 29th Field Conference, p. 231–242.

Shafiqullah, M., Damon, P. E., Lynch, D. J., Reynolds, S. J., Rehrig, W. A., and Raymond, R. H., 1980, K-Ar geochronology and geologic history of southwestern Arizona and adjacent areas, *in* Jenny, J. P., and Stone, C., eds., Studies in western Arizona: Arizona Geological Society Digest, v. 12, p. 201–260.

Shafiqullah, M., Damon, P. E., and Peirce, H. W., 1976, Late Cenozoic tectonic development of Arizona Basin and Range province [abs.]: Proceedings, International Geological Congress, 25th, Aug. 16–25, 1976, Sydney, Australia, v. 1, p. 99.

Sharp, R. P., 1942, Multiple Pleistocene glaciation on San Francisco Mountain, Arizona: Journal of Geology, v. 50, p. 481–503.

Smiley, T. L., 1958, The geology and dating of Sunset Crater, Flagstaff, Arizona, *in* New Mexico Geological Society 9th Field Conference, Guidebook, p. 186–190.

Updike, R. G., and Péwé, T. L., 1974, Glacial and pre-glacial deposits in the San Francisco Mountain area, northern Arizona, *in* Karlstrom, T.N.V., and others, eds., Geology of northern Arizona with notes on archaeology and paleoclimate, Part II, Area studies and field guides: Geological Society of America, Guidebook, Rocky Mountain Section Meeting, Flagstaff, p. 557–566.

Wohletz, K. H., 1978, The eruptive mechanism of the Peridot Mesa vent, San Carlos, Arizona, *in* Burt, D. M., and Péwé, T. L., eds., Guidebook to the geology of central Arizona: Arizona Bureau of Geology and Mineral Technology Special Paper 2, p. 167–171.

MANUSCRIPT ACCEPTED BY THE SOCIETY JANUARY 12, 1985

Indexes

Author Index

Typeset by WESType Publishing Services, Inc., Boulder, Colorado
Printed in U.S.A. by Malloy Lithographing, Inc., Ann Arbor, Michigan